Fire Engineering's Handbook
for Hazardous Materials Response

FIRE ENGINEERING'S HANDBOOK FOR HAZARDOUS MATERIALS RESPONSE

Copyright © 2019 by
Fire Engineering Books & Videos
1421 South Sheridan Road
Tulsa, Oklahoma 74112-6600 USA

Phone: 918.835.3161
Fax: 918.831.9555
800.752.9764
+1.918.831.9421
info@fireengineeringbooks.com
www.fireengineeringbooks.com

Senior Vice President Eric Schlett
Operations Manager Holly Fournier
Sales Manager Josh Neal
Managing Editor Mark Haugh
Production Manager Tony Quinn
Technical Editor Glenn Corbett
Development Editor Chris Barton
Book Designer Sheila Brock
Cover Designer Beth Rose

Printed in the United States of America

1 2 3 4 5   23 22 21 20 19

CONTENTS

4 HAZARDOUS MATERIALS OPERATIONS: MISSION SPECIFIC

Skill Drills

Study Guide

ABOUT THE AUTHORS

RONALD KANTERMAN

is a 44-year fire service veteran and has served various commands at the chief level for the past 30 years. He entered the fire service in 1975 in Brooklyn, New York, and worked as a chief in three subsequent commands in New Jersey and Connecticut.

Kanterman holds a BA in fire service administration and a MS in fire protection management, both from John Jay College in New York City. He holds a second master's degree in environmental sciences from the New Jersey Institute of Technology. He is currently an adjunct professor of fire protection and emergency management at the University of New Haven and has taught for the past 30 years at various two- and four-year colleges at the undergraduate and graduate levels. He also teaches as an adjunct instructor at the National Fire Academy in Emmitsburg, Maryland.

Kanterman has been published more than 180 times and contributes to Fireengineering.com regularly through his blog, "Chief Kanterman's Journal" as well as *Fire Engineering* magazine. In 2019 *Fire Engineering* published his latest book *Fire Officer's Guide to Occupational Safety and Health* and was a contributing chapter author for the inaugural *Fire Engineering's Handbook for Firefighter I & II*, published in 2008. He authored the Health and Safety chapter in the seventh edition of the *Fire Chief's handbook*, published in 2015, and has contributed to six other fire service texts. He published his first book, *Managing Fireworks Displays* in 2008. He sat on the Educational Advisory Committee for and has taught at the Fire Department Instructor's Conference (FDIC) held in Indianapolis for the past 20 years.

Chief Kanterman is an avid volunteer with the National Fallen Firefighters Foundation and worked as the chief of operations for 13 years at the National Memorial Weekend in Emmitsburg, Maryland. He assumed the role of Memorial Weekend Incident Commander in 2018. He is also the lead instructor for the Local Assistance State Teams (LAST) program, which assists fire departments and families who have experienced a line-of-duty-death and is also affiliated with the Firefighter Cancer Support Network.

DAVE PROBO

is the deputy fire marshal and hazardous materials coordinator for Spartanburg County, SC, and adjunct fire instructor for the Hazardous Materials Division of the South Carolina Fire Academy. He retired as a battalion chief and hazardous materials team coordinator for the St. John's Fire Department after 25 years. He has been working in hazardous materials since 1990 and has served as hazardous materials team coordinator since 2000.

INTRODUCTION

Firefighters are much more comfortable responding to and operating at structure fires, rescues, or EMS runs. Therefore, firefighters need to have a heightened awareness of what to do and, moreover, what not to do, in order to avoid injury or worse. Firefighters tend to be less comfortable handling a hazardous materials response.

The focus of this text is to address the responsibilities of firefighters operating at hazmat scenes at the awareness and operational levels and to provide a clear picture about how to handle these types of incidents. The material is covered using the same simple and systematic approach that is employed on scene.

The basic steps include:

- analyzing the incident,

- developing the response plan, and

- implementing the plan.

These concepts can be applied to any hazmat scenario and will ensure an effective and orderly response, especially when multiple response agencies are involved.

The fire service has evolved into an all-hazards organization. Simply put, if it endangers the public, the fire department typically plays some role in the response. In most communities it is the responsibility of the fire department to handle the release of hazardous materials. In most cases these incidents are unintentional and the result of human error or container failure. In other cases, however, we encounter situations where the release is intentional and is classified as a criminal or act of terrorism. In either case, it is most important that firefighters handle the response in accordance with their level of training. Hazardous materials can be found in every community, regardless of its size, in households, commercial and industrial facilities, or in transportation. Even if you believe that no storage or use of these materials exist in your town, there is a good chance it will be transported through your town. Because hazardous materials are used in or transported through every community, proper training prevents unnecessary and potentially dangerous exposures to these materials. Additionally, training is required by OSHA 1910.120 (q), which states that employees who participate—or are expected to participate—in emergency responses should be given training in hazardous materials.

Incidents involving hazardous materials or weapons of mass destruction are different from other emergencies because they can be severe and pervasive with rapidly changing conditions as the materials react with the natural environment or our mitigation methods. While responding to standard fire, rescue, and medical alarms requires a fast response where responders act almost immediately after stepping off the apparatus, hazardous materials incidents require a slower, more methodical approach. Responders must attempt to recognize that it is a hazardous materials incident and identify the product before acting. A release of a hazardous material can endanger responders, the public, the environment, and property if proper actions are not taken to control or stop the release.

Depending on the material, a release can influence a much larger geographical area, which increases the potential for injury or death. Depending on the scale, these incidents can become more complex than standard emergency responses and may require specially equipped and trained personnel to respond. Consider the potential for long-term effects on people, the environment, and property. Incidents involving terrorist activities also bring the additional danger of secondary attacks and armed resistance coupled with the fact that these releases are often designed to inflict the greatest amount of damage or harm. Terrorist incidents also require that crime scene management, evidence preservation, and unified command are employed to work closely with the needs of law enforcement.

NFPA 1072 KNOWLEDGE CORRELATIONS

JPR	FFI Requisite Knowledge	Chapter	Page(s)
	Awareness		
4.1	**Awareness—General**	1	4
4.1.1	Awareness personnel are those persons who, in the course of their normal duties, could encounter an emergency involving hazardous materials/weapons of mass destruction (WMD) and who are expected to recognize the presence of the hazardous materials/WMD, protect themselves, call for trained personnel, and secure the area.	1	4
4.1.3	General knowledge requirements. Role of awareness personnel at a hazardous materials/WMD incident, location and contents of the AJH emergency response plan, and standard operating procedures for awareness personnel.	1	4
4.2.1	Recognize and identify the hazardous materials/WMD and hazards involved in a hazardous materials/WMD incident, and approved reference sources, so that the presence of hazardous materials/WMD is recognized and the materials and their hazards are identified.	1, 2	2, 13, 17
	(A) Requisite Knowledge. What hazardous materials and WMD are; basic hazards associated with classes and divisions; indicators to the presence of hazardous materials including container shapes, NFPA 704 markings, globally harmonized system (GHS) markings, placards, labels, pipeline markings, other transportation markings, shipping papers with emergency response information, and other indicators; accessing information from the Emergency Response Guide (ERG) (current edition) using name of material, UN/NA identification number, placard applied, or container identification charts; and types of hazard information available from the ERG, safety data sheets (SDS), shipping papers with emergency response information, and other approved reference sources.	2	31
	(B) Requisite Skills. Recognizing indicators to the presence of hazardous materials/WMD; identifying hazardous materials/WMD by name, UN/NA identification number, placard applied, or container identification charts; and using the ERG, SDS, shipping papers with emergency response information, an other approved reference sources to identify hazardous materials/WMD and their potential fire, explosion and health hazards.	3, 3	46–48, 57
4.3	**Initiate Protective Actions**		
4.3.1	Isolate the hazard area and deny entry at a hazardous materials/WMD incident, given a hazardous materials/WMD incident, policies and procedures, and approved reference sources, so that the hazard area is isolated and secured, personal safety procedures are followed, hazards are avoided or minimized, and additional people are not exposed to further harm.	2, 3	42, 83–84
4.4.1	Initiate required notifications at a hazardous materials/WMD incident, given a hazardous materials/WMD incident, policies and procedures, and approved communications equipment, so that the notification process is initiated, and the necessary information is communicated.	2	42
	Operations		
5.1	**Operations—General**	1	
5.1.1	Operations level responders are those persons who respond to hazardous materials/weapons of mass destruction incidents for the purpose of implementing or supporting actions to protect nearby persons, the environment, or property from the effects of the release.	1	5, 45
5.1.4	Operations level responders shall have additional competencies that are specific to the response mission and expected tasks as determined by the AHJ.	1	5
5.1.5	General Knowledge Requirements. Role of operations level responders at a hazardous materials/WMD incident; location and contents of AHJ emergency response plan and standard operating procedures for operations level responders, including those response operations for hazardous materials/WMD incidents.	3	46
5.2	**Identify Potential Hazards**		
5.2.1	Identify the scope of the problem at a hazardous materials/WMD incident, given a hazardous materials/WMD incident, an assignment, policies and procedures, and approved reference sources, so that container types, materials, location of any release, and surrounding conditions are identified, hazard information is collected, the potential behavior of a material and its container is identified, and the potential hazards, harm, and outcomes associated with that behavior are identified.	3	46

JPR	FFI Requisite Knowledge	Chapter	Page(s)
5.3.1	Identify the action options for a hazardous materials/WMD incident, given a hazardous materials/WMD incident, an assignment, policies and procedures, approved reference sources, and the scope of the problem, so that response objectives, action options, safety precautions, suitability of approved personal protective equipment (PPE) available, and emergency decontamination needs are identified.	3	53
5.4.1	Perform assigned tasks at a hazardous materials/WMD incident, given a hazardous materials/WMD incident; an assignment with limited potential of contact with hazardous materials/WMD, policies and procedures, the scope of the problem, approved tools, equipment, and PPE, so that protective actions and scene control are established and maintained, on-scene incident command is described, evidence is preserved, approved PPE is selected and used in the proper manner; exposures and personnel are protected; safety procedures are followed; hazards are avoided or minimized; assignments are completed; and gross decontamination of personnel, tools, equipment and PPE is conducted in the field.	3	52
5.5	**Emergency decontamination**	4	
5.5.1	Perform emergency decontamination at a hazardous materials/WMD incident, given a hazardous materials/WMD incident that requires emergency decontamination; an assignment; scope of the problem; policies and procedures; and approved tools, equipment, and PPE for emergency decontamination, so that emergency decontamination needs are identified, approved PPE is selected and used, exposures and personnel are protected, safety procedures are followed, hazards are avoided or minimized, emergency decontamination is set up and implemented, and victims and responders are decontaminated.	4	108
5.6	**Progress Evaluation and Reporting**	3	86
5.6.1	Evaluate and report the progress of the assigned tasks for a hazardous materials/WMD incident, given a hazardous materials/WMD incident, an assignment, policies and procedures, status of assigned tasks, and approved communication tools and equipment, so that the effectiveness of the assigned task is evaluated and communicated to the supervisor, who can adjust the Incident Action Plan (IAP) as needed.	3	86

Operations Mission-Specific

JPR	FFI Requisite Knowledge	Chapter	Page(s)
6.1	**Operations Mission-Specific—General**	4	96
6.1.1	Operations level responders assigned mission-specific responsibilities at hazardous materials/WMD incidents are those operations level responders designated by the AHJ to perform additional tasks to support the AHJs response mission, expected tasks, equipment, and training in the following areas:	4	96
	(1) Personal protection equipment (see Section 6.2)		
	(2) Mass decontamination (see Section 6.3)		
	(3) Technical decontamination (see Section 6.4)		
	(4) Evidence preservation and sampling (see Section 6.5)		
	(5) Product control (see Section 6.6)		
	(6) Detection, monitoring, and public safety sampling (see Section 6.7)		
	(7) Victim rescue and recovery (see Section 6.8)		
	(8) Illicit laboratory incidents (see Section 6.9)		
6.2	**Personal Protective Equipment**	3, 4	79, 98
6.2.1	Select, don, work in, and doff approved PPE at a hazardous materials/WMD incident, given a hazardous materials/WMD incident; a mission-specific assignment in an IAP that requires use of PPE; the scope of the problem; response objectives and options for the incident; access to a hazardous materials technician, an allied professional, an emergency response plan, or standard operations procedures; approved PPE; and policies and procedures, so that approved PPE is selected, inspected, donned, worked in, decontaminated, and doffed; exposures and personnel are protected; safety procedures are followed; hazards are avoided or minimized; and all reports and documentation pertaining to PPE use are completed.	4, 4	98, 105
6.3	**Mass Decontamination**	4	107– 108, 114–116
6.4	**Technical Decontamination**	4	107–109, 112
6.5	**Evidence Preservation and Public Safety Sampling**	4	116–120

Introduction to Hazardous Materials/WMD

This chapter provides knowledge and information related to the following NFPA Standard 1072 JPRs.

4.1.1	5.1.5
4.1.3	
5.1.1	
5.1.4	

OBJECTIVES

Upon completion of the chapter, you should be able to do the following:

- Describe the difference between hazmat/weapons of mass destruction (WMD) incidents and other emergencies.
- Identify different locations through which hazardous materials are either stored or shipped.
- Describe the capabilities and limitations of the awareness and operations-level responder at a hazmat/WMD incident.
- Describe the impact hazardous materials laws and regulations have on emergency responses.
- Describe how the local emergency response plan (LERP) and fire department standard operating procedures (SOPs) guide the operations at a hazmat/WMD incident.
- Define commonly used hazardous materials terminology as covered in this chapter.
- List the three main phases of a hazmat/WMD response.

INTRODUCTION

The fire service has evolved over the years into an all-hazards organization. Simply put, if it endangers the public, the fire department typically plays some role in the response. Along those lines, in most communities it is the responsibility of the fire department to handle the release of hazardous materials (fig. 1–1). In most cases these incidents are unintentional and the result of human error or container failure. In some cases, however, we encounter situations where the release is intentional and is classified as a criminal or terrorist act. In either case, it is important that firefighters handle the response in accordance with their level of training. Because hazardous materials are used in or transported through every community, proper training prevents unnecessary and potentially dangerous exposures to these materials (fig. 1–2). Additionally, training is required by OSHA 1910.120 (q), which states that employees who participate—or are expected to participate—in emergency responses be given training in hazardous materials.

Fig. 1–1. Fire departments are expected to handle all types of emergencies, including those involving hazardous materials. (Courtesy Tom Lenart)

Fig. 1–2. Hazardous materials are transported through every community in many types of vessels and packages. Note the tanker transporting corrosives above. (Courtesy Emergency Training Solutions)

Historically, firefighters tend to be less comfortable handling a hazardous materials (hazmat) response rather than structural fires, rescues, or EMS runs, partially because of the potential complexity of these incidents but also due to the catastrophic incidents which have occurred throughout the world. The focus of this text is to address the responsibilities of firefighters operating at the hazmat scenes at the awareness and operations levels and to provide a clear picture about how to handle these types of incidents. The material is covered using the same simple and systematic approach that is employed on scene. The basic steps include analyzing the incident, developing the response plan, and then implementing the plan. These concepts can be applied to any hazmat scenario and will ensure an effective and orderly response, especially when multiple response agencies are involved.

HAZMAT/WMD INCIDENTS VS. OTHER EMERGENCIES

Hazardous materials/weapons of mass destruction incidents are different from other emergencies because they can be far more pervasive and severe than, say, a large structure fire. Responding to standard fire, rescue and medical alarms requires a fast response where responders act almost immediately after stepping off the apparatus. Hazardous materials incidents require a slower more methodical approach. Responders will attempt to recognize that it is a hazardous materials incident and identify the product before taking action. A release of a hazardous material can endanger responders and the public as well as the environment and property if proper actions are not taken to control or stop the release. Also, depending on the material, a release can affect a much larger geographical area, which increases the potential for injury or death. Depending on the scale, these incidents can become more complex than standard emergency responses and may require specially equipped and trained personnel to respond (fig. 1–3). The potential for long-term effects on people, the environment, and property should also be considered. Incidents involving terrorist activity also bring the additional danger of secondary attacks and armed resistance coupled with the fact that these releases are often designed to inflict the greatest amount of damage or harm (fig. 1–4). Terrorist incidents also require that crime scene management, evidence preservation, and unified command be employed to work closely with the needs of law enforcement.

Fig. 1–3. Hazmat incidents are often more complicated than standard emergency responses. (Courtesy U.S. Air Force)

Fig. 1–4. Terrorist incidents such as the 1995 bombing of the Alfred Murrah Federal Building in Oklahoma City are designed to inflict great damage, widespread injury and death. (Courtesy U.S. Department of Defense)

Fig. 1–5. Some communities have facilities that manufacture hazardous materials.

Fig. 1–6. Hazardous materials are transported by rail through some communities. More often, they are transported by tank truck or common carrier. (Courtesy Emergency Training Solutions)

COMMON HAZARDOUS MATERIALS LOCATIONS

While we know that hazardous materials are used, stored, and transported in every community, it is important that responders recognize both the obvious and not-so-obvious locations where these materials may be found. Some response districts have facilities from which hazardous materials are produced and shipped (fig. 1–5). Others may use the materials in manufacturing facilities, and some may deal only in consumer quantities and vehicles that travel on their roadways and railways (fig. 1–6). It is important to understand that no matter how small or large the community, it is always possible to encounter a hazardous material.

Response personnel should conduct pre-emergency plans of those facilities in their response area that utilize hazardous materials in quantities that could be harmful if released. Chapters 2 and 3 discuss fixed facilities in depth. Table 1–1 lists some of the locations where responders should expect to see hazardous materials.

Introduction to Hazardous Materials/WMD

HAZMAT

Chapter 1

Table 1–1. Locations where hazardous materials are found

Methods of Transportation	
Roadways	Railways
Waterways	Pipelines
Airways	

Fixed Facility Locations	
Large manufacturing or storage plants	Photo processing laboratories
Service stations	Fuel storage facilities
Doctors' offices	Agricultural stores
Hardware stores	Tank farms
Dry cleaners	Weapons depots
Paint supply stores	Warehouses
School laboratories	Laboratories
Farms	Maintenance facilities
Residences	Flight line areas
Pool supply stores	

Transportation Terminal Locations	
Docks or piers	Airplane hangers
Railroad stations	Truck terminals

HAZMAT RESPONDER LEVELS AND RESPONSIBILITIES

It is important to remember that responders must operate within their level of training and never above it. You need to understand what you can't do vs. what you can do. To do so would put both first responders and the public at risk. The **National Fire Protection Association (NFPA)** and the **Occupational Safety and Health Administration (OSHA)** both recognize the following levels of training and responsibility as they relate to hazardous materials incidents.

Awareness-level responder

HM 4.1.1 Awareness-level responders are expected to recognize the presence of hazardous materials, secure the area, protect themselves and others, and request the response of appropriately trained personnel to the scene (fig. 1–7). These responders include those individuals who in the course of their work, could encounter a hazardous materials incident. This group includes

public works, utilities workers, sanitation workers, and security personnel. Awareness-level responders do not typically train to play an active role in offensive or defensive operations. They recognize and identify hazardous materials and/or hazardous materials incidents, initiate protective actions by isolation of the incident, and deny entry to unauthorized persons to prevent others from further harm. They make notification by requesting the appropriate assistance needed such as fire, EMS, and law enforcement. These actions need to be performed in a manner consistent with departmental SOPs/SOGs and the local emergency response plan (LERP). Chapter 2 covers the awareness level tasks in depth along with the job performance requirements (JPR's) as set forth in NFPA 1072, *Standard for Hazardous Materials/Weapons of Mass Destruction Emergency Response Personnel Professional Qualifications.*

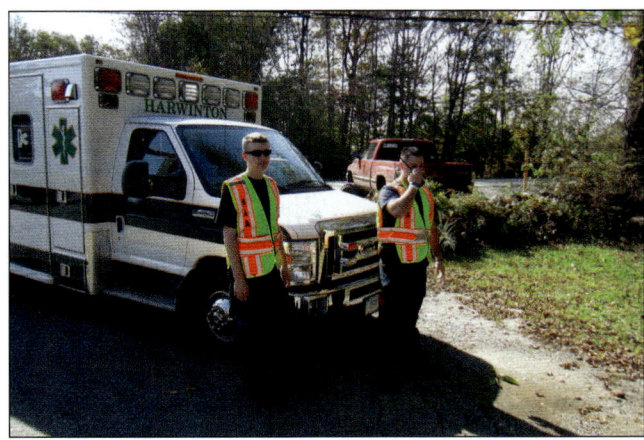

Fig. 1–7. EMS personnel typically operate at the awareness level.

Operations-level responder

HM 3.4 Operations-level responders operate above the awareness level and in the course of their normal job duties respond to hazardous materials emergencies or potential hazardous materials emergencies in order to protect people, the environment, and property from the effects of the release. Typically, this includes personnel assigned to fire service, law enforcement and EMS. This response is carried out by using a defensive approach to control the release from a safe distance and to prevent it from spreading. Operations-level responders are trained to act in a defensive manner only, without attempting to stop the release of the material at its source unless it can be done from a remote location (fig. 1–8).

HM 4.1.3 The role of the personnel operating at this level is based on the department's SOPs/SOGs as well as the community's LERP.

Fig. 1–8. Operations-level personnel can operate in a defensive manner only and not take an offensive approach to stopping the release.

Although hazmat responses do not typically make up the bulk of emergency calls for a fire department, hazardous materials are used in or transported through every community, which means there is always an opportunity to respond to that type of incident. Operations-level training prevents unnecessary and potentially dangerous exposures to response personnel and civilians alike. OSHA 29 CFR 1910.120 (q) requires this level of training for responders conducting defensive operations at a hazmat incident.

HM 5.1.1 The responsibilities of operations-level responders include identify potential hazards, identifying action options, and then implementing the action plan. Also, as part of the action plan to protect the public, operations-level responders shall perform gross and emergency decontamination. In addition to the duties assigned to operations-level responders, authorities having jurisdiction (AHJ) can elect to have responders perform mission specific tasks. Per the NFPA#1072 standard (2017 ed.), operations-level responders must receive additional training to perform any of the following mission specific operations: personal protective equipment (PPE), mass decontamination, technical decontamination, evidence preservation and sampling, product control, air monitoring and sampling, victim rescue and recovery, and responses to illicit laboratory incidents. Chapter 4 covers the operations-level tasks in depth along with the job performance requirements (JPRs) as set forth in NFPA#1072, *Standard for Hazardous Materials/Weapons of Mass Destruction Emergency Response Personnel Professional Qualifications.*

Technician-level responder

Technician-level responders operate above the awareness and operations level in the course of their normal job duties responding to hazardous materials emergencies or potential hazardous materials emergencies in order to protect people, the environment, and property from the effects of the release (fig. 1–9). Typically, this includes personnel assigned to fire service, law enforcement and EMS. This response is carried out using an offensive approach to mitigate the release to prevent it from spreading and ensuring it is no longer a threat to the public.

Fig. 1–9. Technician-level responders operate with a higher level of training and equipment. (Courtesy U.S. Navy)

Hazardous Materials Specialist

HM 5.1.4 Hazmat materials specialists are persons with expert knowledge and training to assist with the mitigation of the emergency. Some specialists are product specialist and have extensive knowledge with the hazardous material involved. Usually these people come from chemical manufacturers or partnerships with universities. There are also container specialists, who have extensive knowledge of the over the road containers, railcars or intermodals. Usually these personnel come from the transportation corporations such as clean up contractors, trucking companies and the railroad.

HAZARDOUS MATERIALS LAWS AND REGULATIONS

There are numerous federal laws that are related to hazardous materials response, planning, and operations. It is key that the emergency response community understands these laws and the requirements set for them. Federal laws are created by the U.S. Congress, while rules and regulations are created by federal or other governing bodies to provide for ease of interpretation and compliance. The following is a synopsis of the laws and their basic requirements.

OSHA HAZWOPER

In states that have OSHA plans, the **Hazardous Waste Operations and Emergency Response (HAZWOPER; 29CFR1910.120)** regulations are enforced by OSHA compliance officers. In the remaining states, these same regulations have been adopted by the federal **Environmental Protection Agency (EPA)** under 40CFR311. This law with its accompanying regulations applies to all emergency responders whether public or private and includes hazmat teams, police officers, firefighters, and industrial emergency responders. For emergency responders, this law contains the following levels of competency: awareness, operational, technician, specialist, and on-scene incident commander. It also includes requirements for skilled support personnel (e.g., heavy equipment operators) and specialist employees (e.g., an on-site scientist in a research facility). In fixed facilities, one may find emergency response teams that are trained to the various levels of competency along with skilled support personnel and specialist employees. During pre-incident planning of fixed facilities, it is imperative that fire departments, hazmat teams, and other emergency response agencies seek out these groups and understand their roles during a site emergency.

Comprehensive Environmental Response Compensation and Liability Act

The **Comprehensive Environmental Response Compensation and Liability Act (CERCLA)**, also known as the Superfund, covers unattended, abandoned, or inactive hazardous waste disposal sites. This law also requires that the "responsible party" (who caused the hazmat release) be held liable and must contact the National Response Center upon release of a "reportable quantity."

Superfund Amendments and Reauthorization Act (SARA)

The **Superfund Amendments and Reauthorization Act (SARA-1986)** amended CERCLA and also added sections on planning, preparedness, training, and response, which in turn had the most influence on "the way we do business" of most laws pertaining to hazmat. This law, in turn, led to the development of OSHA 1910.120 HAZWOPER. In addition, SARA Title III (Emergency Planning and Community Right to Know Act) led to the development of state and local emergency planning committees known as state emergency response commissions (SERCs) and local emergency planning committees (LEPCs), respectively.

The **Resource Conservation and Recovery Act (RCRA)** requires the proper management and disposal of hazardous waste in treatment, storage, and disposal facilities and establishes installation, leak prevention, and notification requirements for underground storage tanks. Most large industrial facilities have storage areas for their waste, known as RCRA pads, which may or may not have fire protection.

Clean Air Act

The **Clean Air Act (CAA)** was enacted to control airborne emissions to protect the environment. When amended in 1990, it addressed emergency response planning at certain facilities that produce highly hazardous or toxic materials. In addition, it authorized the creation of the Chemical Safety and Hazard Investigation Board and operates similarly to the **National Transportation Safety Board (NTSB)**.

Oil Pollution Act

The **Oil Pollution Act (OPA)** amended the original federal Water Pollution Control Act. It covers fixed facilities and carriers (barges, tankers, railcars, deep-water terminals, vessels, pipelines) and requires emergency plans, drills, exercises, and the verification of spill response contractors and their capabilities. Where a coastal zone may be affected, area committees and **area contingency plans (ACPs)** are required.

Community Emergency Planning Regulations (40CFR 301–303)

These regulations were developed for SARA Title III, and they require that state and local emergency planning groups develop and/or review hazardous materials response plans (fig. 1–10). The SERCs do this on the state level, and the LEPCs do it on a local level. LEPCs typically include the local fire department, law enforcement, emergency medical service, government representatives, health department, nonprofit assistance agencies, and perhaps local industry.

Fig. 1–10. Local officials and emergency responders both play a vital role in the LEPC.

HAZARDOUS MATERIALS GENERAL RESPONSE GUIDELINES

Two valuable resources for hazmat responders are the LERP and department SOPs/SOGs. These documents can provide valuable guidance on how to handle a hazardous materials release in the community.

Local Emergency Response Plan (LERP)

A good place for the responder to learn about hazardous materials/WMD response is the local emergency response plan, which are found in most, if not all, communities. Many times, an annex specific to hazardous materials is included in that plan. Plans for terrorism events are also are also annexed in LERP's as the personnel responding

to these types of incidents must have the same training as a hazardous materials responder.

The LERP addresses how the local response will occur and if necessary, how that response becomes elevated to the state and even the federal level. The LERP is usually maintained in the local communications center, the 9-1-1 center and/or the emergency operations center (EOC) in each community. A copy should be located in each fire station so that department members can become familiar with its contents.

Standard operating procedures

Progressive departments have SOPs or SOGs for all types of department operations, including hazardous materials responses. These documents are designed to give responders guidance on how to handle different types of incidents they may encounter in the field. Responders should be familiar with, and know how to access their department's SOPs or SOGs. Knowing how your department operates is the key to safe and successful outcomes. A copy of these documents should be provided in each fire station.

COMMON TERMINOLOGY

It is important that hazmat responders be familiar with basic terms that are referenced in hazmat standards and responses. This familiarity enables responders from various backgrounds to speak the same "language" at the scene. It is important to understand that although all of these materials are classified as hazardous materials, there are several terms used that are based on which agency created the definition. While it may seem confusing, taking the time to understand the language will pay large dividends.

Hazardous materials definitions

A **hazardous material** is defined by the U.S. Department of Transportation (DOT) as any substance or material capable of posing an unreasonable risk to health, safety, and property when transported in commerce, and which has been so designated. In Canada, the term used for these materials is **dangerous goods**.

Hazardous substance

A **hazardous substance** is a substance designated under the CERCLA as posing a threat to waterways and the environment if released. In OSHA 29 CFR 1910.120 the term is used to describe all the chemicals that are regulated by the EPA as hazardous substances and by the DOT as hazardous materials.

Hazardous chemical

A **hazardous chemical** is defined in OSHA 29 CFR 1910.120 as any chemical that poses a physical or health hazard.

Dangerous goods

Dangerous goods, as defined by the Canadian Transport Commission (CTC), describes any product, substance, or organism included by its nature or by the regulation in any of the United Nations (UN) classes of hazardous materials.

Hazardous waste

A **hazardous waste**, as defined by the EPA in the RCRA (Resource Conservation and Recovery Act), is waste that poses a serious to persons and/or the environment.

Weapons of mass destruction

A **weapon of mass destruction (WMD)** is a material or device designed to inflict great harm on people or property. It can be an explosive device, a toxic or poisonous chemical, a disease organism, or radioactive device (fig. 1–11).

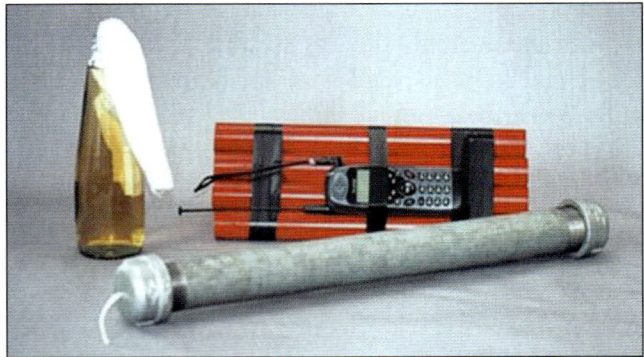

Fig. 1–11. Improvised explosive devices (IEDs) are considered weapons of mass destruction. Shown here are a Molotov cocktail, explosives with a cell phone as the actuator, and homemade pipe bomb.

MANAGING HAZMAT/WMD INCIDENTS

Due to the potential complexity of a hazmat/WMD incident, it is important that responders follow a logical and standardized course of action to ensure that all response objectives are met and the safety of the responders and civilians alike is addressed. The following steps are used to guide response personnel through the response.

Analyze the incident

Like a standard emergency response, this phase begins the moment the emergency is dispatched. The information given to the responder from the dispatcher on the initial assignment, as well as updates given en route, will be used to make initial decisions about how to approach the scene (fig. 1–12). Information such as the location and if possible, the materials involved provides a solid base to build upon. Additional information collected will be used to further analyze the incident and to plan and implement the response. The key steps in this process are to survey the scene, collect hazard and response information, and determine the likely behavior of the material. During this response phase, personnel isolate the area and deny entry to prevent any additional exposures as they collect information (fig. 1–13).

Fig. 1–12. Size-up begins with the information provided by the dispatcher. (Courtesy J. Rains)

Fig. 1–13. The collection of information is used to plan the appropriate response.

Planning the response

After the data on the material(s) involved is collected and analyzed, it is used to plan the proper response. It may be determined that the resources currently on scene are appropriate or that more advanced equipment and trained personnel are required. Issues such as scene control and safety are considered and weighed, based on the material(s) involved. The incident command system positions are filled in order to support the options chosen for the plan. The response objectives are defined along with the means by which to accomplish those objectives, including selection of PPE (fig. 1–14).

Fig. 1–14. The information collected in the initial phase of the response is used to determine the level of PPE required.

Implementing the response

During this phase, the chosen response objectives are implemented and product control or rescue operations are conducted (fig. 1–15). This phase should be continuously evaluated to ensure that the objectives are being met, and adjustments made if needed. Personnel should be decontaminated at the end of this phase to ensure that there is no potential for the material to spread outside of the containment area. Decontamination is discussed in depth in chapter 4.

Fig. 1–15. The response objectives are implemented during the response phase of the operation.

SAFETY CONSIDERATIONS

Responder safety is of paramount importance. The difference between fires and hazmat incidents, more often than not, is there's more time at a hazmat emergency to analyze the scene and plan and execute, unlike a building fire with persons trapped inside. A fire emergency normally requires split second decision making and quick action. Part and parcel to planning a hazmat response is to have a safety plan specific to the response and a safety officer on the scene to administer the plan. All responders must be connected to the plan and understand their tactical role. It is important that everyone knows the goals and objectives of the operation, who will be doing which task, emergency communications and procedures, properties of the material(s) involved and having a true respect for the materials involved. All too often, personnel get too comfortable at a scene which could lead to injury, illness or death. All aspects of a hazmat response revolve around scene safety. Some of these are PPE selection, understanding basic chemistry, product identification, quantity of the product involved, respiratory protection, understanding toxicology, the principles of decontamination, incident management and scene control, monitoring, and physicals hazards.

SUMMARY

Today's fire service includes hazardous materials incident response within the many types of incidents we of which we are responsible. These incidents are often more involved than a standard emergency response and require special training and equipment. Hazardous materials can be found in any community, regardless of its size, in households, commercial and industry facilities, or in transportation. It's important to note that even if there is no storage or use of these materials in your town, there is a good chance it will be transported through your town. Also remember that most if not all jurisdictions have at least one gas station for fueling motor vehicles and an average sized tank truck, which delivers 9,500 gallons of gasoline. There are many laws that govern the use and accidental release of hazardous materials to which responders can look for guidance. Additionally, the LERP or department SOPs should provide additional direction and guidance to emergency responders.

QUESTIONS

1. Which training level has the primary duties of protecting others from harm and requesting assistance?

2. How are hazmat incidents different from other emergencies?

3. What type of hazardous materials location would a photo processor be?

4. What are the three levels of training recognized by NFPA and OSHA related to hazardous materials incidents?

5. Which NFPA Standard applies to a hazmat response?

6. Name three federal laws that relate to a hazardous materials response.

7. Community Emergency Planning Regulations call for each state to form what type of group, and what is the group's primary responsibility?

8. What are the three steps to ensuring all hazmat response objectives are met?

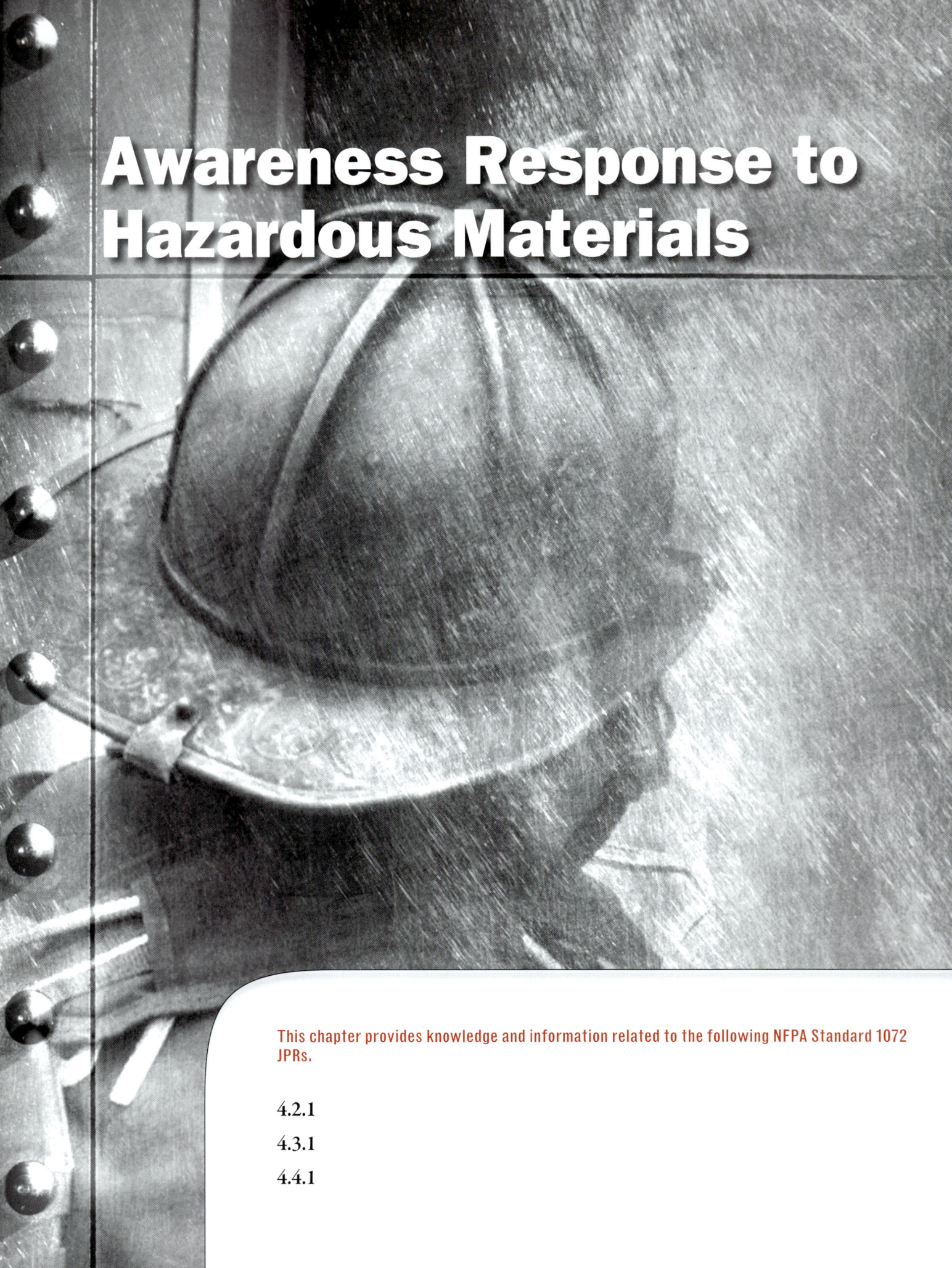

Awareness Response to Hazardous Materials

This chapter provides knowledge and information related to the following NFPA Standard 1072 JPRs.

4.2.1

4.3.1

4.4.1

OBJECTIVES

Upon completion of the chapter, you should be able to do the following:

- Recognize the presence of hazardous materials.

- Have knowledge of the written resources to help identify materials and their properties, e.g., shipping papers, waybills, hazardous waste manifest, bill of lading.

- Understand the different labeling systems, e.g., NFPA 704, Global Harmonization labeling, HMMS, military markings, and CAS numbering system.

- Understand the Federal Department of Transportation (DOT) placard system.

- Understand the use of the *Emergency Response Guidebook*.

- Describe procedures for isolating incidents and denying entry.

- Describe procedures for notifications at hazardous materials incidents.

INTRODUCTION

After the product and hazard information has been collected, the responder must then take that information and, combined with assessing the potential behavior of the material, determine the proper course of action. From there, decisions are made regarding scene control, the incident management system, response objectives, and PPE. For incidents involving terrorist activities or WMD, additional concerns must be integrated in the planning and decision-making model. The first step in this process is to recognize the presence of a hazardous material or incident.

RECOGNITION AND IDENTIFICATION

Locations

HM 4.2.1 Hazardous materials can be found in many locations in communities regardless of size. Generally hazardous materials are found in fixed facilities that are either manufacturing plants or storage and use facilities. Also, hazardous materials are transported north, south, east and west by road, rail, sea and air.

Fixed facilities. The term **fixed facility** means a permanent installation as opposed to a mobile operation. Common large fixed facilities that contain hazardous materials are bulk oil storage terminals, refineries, and other major industrial sites. It is important to note, however, that there are many more smaller facilities in every community that store, use, and dispose of hazardous materials. Table 2–1 shows a partial listing of the most common types of fixed facilities around the country and some of the hazards associated with them. Ask yourself if you have any of these in your community, and if so, do you have pre-emergency plans and a good understanding of what is stored and how to manage it? These are crucial elements for the safety of the community and safety of firefighters and other emergency responders. It is important that emergency responders establish a relationship with the fixed facilities in their jurisdiction and conduct frequent facility tours and preplans. This will allow responders to understand the hazardous materials that are present, safety systems in place and have input in facility emergency response plans. Also, by establishing a partnership with these facilities they may provide fire departments with necessary equipment and training to enhance their response capabilities.

Table 2–1. Fixed facilities with hazardous materials

Type of Fixed Facility	Hazards
Gasoline station	Gasoline, diesel fuel, motor oils, degreasers
Home center	Lumber, paints, pesticides, propane, flammable, and combustible liquids, adhesives
Lumberyard	Bulk lumber storage, paints, solvents, adhesives
Bulk fuel depot	Flammable and combustible liquids
Refinery	Flammable and combustible liquids, flammable gas
Specialty chemicals	Various hazardous materials of all classes
Hardware store	Paints, flammable and combustible liquids, adhesives
Dry cleaning plants	Flammable and toxic materials
Garden centers	Pesticides, herbicides, flammable and combustible liquids
Photo processing labs	Flammable and toxic materials

Type of Fixed Facility	Hazards
Farm supply stores	Pesticides, herbicides, flammable and combustible liquids
High schools/ colleges (labs)	Flammable and combustible liquids, acids, corrosives, poisons, radioactive materials, water-reactives, oxidizers, organic peroxides, compressed gases
Research labs	Flammable and combustible liquids, acids, corrosives, poisons, radioactive materials, water-reactives, oxidizers, organic peroxides, compressed gases, strong magnetic fields
Bulk pharmaceutical plants	Flammable and combustible liquids, acids, corrosives, poisons, radioactive materials, water-reactives, oxidizers, organic peroxides, compressed gases
Water treatment plants	Bulk storage of chlorine
Hospitals	Flammable and combustible liquids, acids, corrosives, poisons, radioactive materials, water-reactives, oxidizers, organic peroxides, compressed gases, strong magnetic fields
Grain silos	Entrapment/suffocation hazard, explosion hazard (dust)

FACILITY AND TRANSPORTATION MARKINGS

Container markings can play a key role in identifying the hazards of a particular product, and potentially even the identity of the material involved. The goal is to identify these markings as early in the incident as possible so that the firefighter can make informed decisions about how the release should be handled. These markings should be identified from a distance to ensure that the responder stays out of the exposure area. Sometimes, in addition to placarding, the container itself will be marked with the name of the material that it holds (fig. 2–1).

Fig. 2–1. Container markings play a key role in product identification.

Facility markings

Various types of markings can be found on hazardous materials containers in fixed facilities. They can range from placarding that indicates the general hazards of the substance to the name of the product listed on the tank. The following are different types of markings found on fixed facility containers.

National Fire Protection Association (NFPA) 704. NFPA standard 704 is a marking system used for fixed bulk storage tanks and vessels in fixed facilities. Many state and local fire codes require that the NFPA 704 system be used at all fixed facility locations within their jurisdiction. It has four diamond-shaped sections joined to create one large diamond placard-type marking (fig. 2–2). This placard system provides information about the general hazards and degree of severity for flammability, toxicity, and reactivity.

Fig. 2–2. The NFPA 704 marking system consists of color-coded diamonds with a number system to indicate the level of danger for specific hazards.

The numbering system ranges from 0 to 4, 0 being no hazard and 4 being the worst hazard. The left diamond section (blue) denotes the health hazard of the material; the upper diamond (red) denotes the flammability of the material; the right diamond section (yellow) denotes the reactivity of the material; and the bottom diamond section (white) denotes any special hazard associated with the material.

Tables 2–2 through 2–5 illustrate the hazards associated with each number given in the different sections of the diamond. This system does not identify the product, so identification must be made through other means. Also, this system does not provide specific information on the hazards, but it only provides a general estimate of how dangerous the hazard is in the context of each of the categories.

Table 2–2

Health Hazard			
Symbol/ Number	Level	Danger	Definition
4	Extreme	Highly toxic	Short-term exposure may be fatal. Special protective equipment may be required.
3	Serious	Toxic	Avoid inhalation or skin contact.
2	Moderate	Moderately toxic	May be harmful if inhaled or absorbed through the skin
1	Slight	Slightly toxic	May cause slight irritation
0	Minimal		All chemicals have some degree of toxicity

Table 2–3

Flammability			
Symbol/ Number	Level	Danger	Definition
4	Extreme	Extremely flammable gas or liquid	Flash point below 73°F
3	Serious	Flammable	Flash point from 73°F to 100°F
2	Moderate	Combustible	Requires moderate heating to ignite. Flash point 100°F to 200°F
1	Slight	Slightly combustible	Requires strong heating to ignite
0	Minimal		Will not burn under normal conditions

Table 2–4

Reactivity Hazards		
Symbol/ Number	Level	Definition
4	Extreme	Explosive at room temperature
3	Serious	May explode if shocked or heated under confinement or mixed with water
2	Moderate	Unstable, may react with water
1	Slight	May react if heated or mixed with water
0	Minimal	Normally stable, does not react with water

Table 2–5

Special Hazard Symbols	
Symbol/ Number	Definition
W	Water reactive
OX	Oxidizer
COR	Corrosive
☢	Radioactive

Hazardous Materials Identification System (HMIS). The HMIS provides clear, recognizable information to facility employees by standardizing the presentation of chemical information. This system was designed to inform workers of the dangers posed by exposure to hazardous materials they work with under normal conditions. This is accomplished with color codes corresponding to the hazards of a product, assigned numeric ratings indicating the degree of hazard, and alphabetical codes designating appropriate personal protective equipment (PPE) employees should wear while handling the material.

In many respects the HMIS is similar to the NFPA 704 marking system. The color and number coding are identical for fire, health, and reactivity hazards. But instead of the diamond, the HMIS uses a color bar system (fig. 2–3). In this system, the white section is used to indicate what level of protective equipment is required instead of the special hazard denoted by white in the 704 system. This is indicated by a letter, with each letter specifying a different level of protection. Another feature that differs from the NFPA label system is that HMIS allows an asterisk (*) to designate a material as a carcinogen (cancer-causing material) or for materials known to have an adverse effect given chronic exposure.

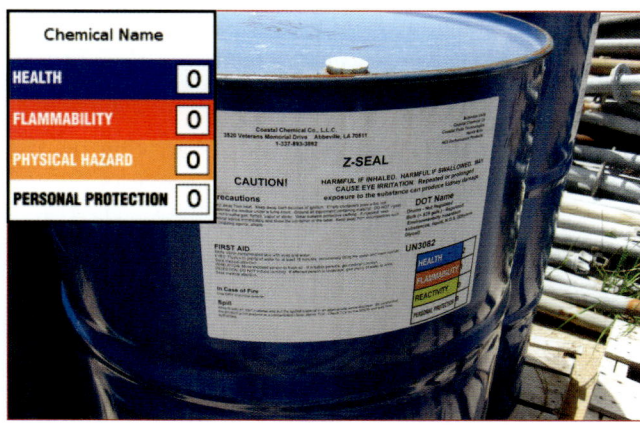

Miscellaneous container markings—fixed facility.

In addition to the placards and symbols discussed previously, the first responder will often encounter other markings on fixed facility tanks that can help determine the contents of the container and provide other valuable information. For example, the product name and/or the capacity of the tank may be labeled or stenciled directly on the tank. Another helpful piece of information is the tank or site identification number (fig. 2–4). This ID number can be used in conjunction with site preplans or emergency operations plans to determine the contents of the tank as well as its volume.

Fig. 2–4. Site identification numbers or the material name stenciled on the tank aids in the identification process.

Transportation markings

Transportation markings can be found on all modes of transport and can include placarding indicating the hazard class to which the material belongs, pipeline markings, package labeling, and other container markings. These markings can aid in understanding the general hazards associated with the loads as well as more specific information, depending on how well it is marked.

Placarding. The Department of Transportation's hazard classification system can provide valuable response information at a hazmat incident. The placards required to be displayed on transportation containers allow emergency response personnel to recognize potential response hazards early in the incident and before more specific information can be collected.

DOT hazards and divisions. The U.S. DOT sets placarding requirements for hazardous materials being transported from one location to another. Materials are placarded according to their primary hazard based on the UN hazard classification system. Responders should be aware, however, that a material grouped in one class may have other hazards associated with it that are not immediately apparent because there is no placard for it. The classifications are broken up into nine classes, with several having subcategories or divisions to account for specific issues. For example, Class 2 is for gases, but there is a division for flammable, nonflammable, poisonous, and so on. There is no division for refrigerated liquefied gases and cryogenic liquids, and they are placarded based on the primary hazard of the substance. These materials have boiling points colder than –90°C (–130°F) at atmospheric pressure.

Table 2–6 illustrates each hazard class, the divisions that make up that class, the representative placard, the hazard, and examples of materials found in the particular class/division.

Placarding requirements. The U.S. DOT requires that hazmat loads of 1,001 lb (454 kg) or more being transported by highway or rail be marked with the appropriate placard indicating the material's primary hazard. The placards must be visible on all four sides of the vehicle. The following materials, however, must be placarded if they are transported in any quantity and are listed as 49CFR Section 172.504 Table 1 and include the following hazard class and divisions:

- 1.1, 1.2, 1.3 explosives
- 2.3 poisonous gas
- 4.3 dangerous when wet
- 5.2 organic peroxide (type B, liquid or solid, temperature controlled)
- 6.1 poison (material poisonous by inhalation)
- 7 radioactive (radioactive yellow III)
- All other materials are placarded under 49CFR Section 172.504 Table 2. These items are only

required to be placarded if there is more than 1,000 lbs (454 kg). If there is a load of hazardous materials that has a combined load of more than 1,000 lbs (454 kg) but no one material exceeds 1,000 lbs (454 kg) then it must be placarded with a *"dangerous"* placard (fig. 2–5). For instance, if the load is 500 lbs (227 kg) of oxidizer, 500 lbs (227 kg) of corrosive and 500 lbs (227 kg) of poison this would carry a *"dangerous"* placard. With a dangerous placard, responders will have no way of determining what hazards are present.

Fig. 2–5. With the dangerous placard, identification of the specific hazard is not readily apparent.

Non-bulk shipments (not exceeding 1,000 lb [454 kg]) of two or more categories of hazardous materials that require placards may use the individual placards for each material instead of a *"dangerous"* placard. The placarding rules require that if 1,000 lb (454 kg) or more of one category of material is loaded at one facility, then the placard for that specific material must be used.

An additional category, *other regulated materials* (ORM-D), are consumer commodities that present a limited hazard during transportation due to form, quantity, or packaging. A placard is not used, only labels (fig. 2–6).

Fig. 2–6. Other regulated materials (ORM-D) label

A consumer commodity is a material that is packaged or distributed in a form intended or suitable for sale through retail sales agencies, or a material used by individuals for personal care or household use. This term also includes drugs and medicines as well as small arms ammunition and furniture polish.

Limitations of placards. Responders arriving on the scene of an incident involving hazardous materials need to be aware that there are limitations to the placarding system. The first thing to remember is that only the primary hazard is placarded, and there may be other dangers associated with the material not indicated by the placard. If the vehicle or tanker was involved in a collision or engulfed in flames, the placards may not be visible. In some instances, response personnel may not be able to get close enough to see the placards. Human error can always play a part as well. Some shipments may be unmarked or improperly marked, and hazards may not be immediately apparent. For example, if the placards were never applied, applied incorrectly, or if they were never changed from the last load hauled in the container, the responder will have difficulty determining the hazard associated with the load. In the case of a dangerous placard, the primary hazard associated with the load may not be apparent.

Another limitation to consider is the fact that there is essentially no difference in danger between a 1,000-lb (454-kg) load of a hazardous substance and a 1,001-lb (454-kg) load of the same substance, even though the first one doesn't need a placard and the second one does. Responders must understand there may be instances in which they encounter unclassified materials or materials with exemptions. While collecting response information, always cautiously approach containers with shapes that commonly indicate hazardous materials.

United Nations Numbering System (UN). This international system of identifying hazardous materials and dangerous goods has been adopted by most countries around the world. The four-digit UN numbers are required for shipment of hazardous materials. The DOT regulations require all vehicles to apply placards using this numbering system as well as the hazard class. North American (NA) numbers are identical to UN numbers and can be assigned to a material if it does not have a UN number.

There is a two-part system to UN/NA identification numbers: the first part divides hazardous materials into nine hazard classes that form the basis for the DOT hazard classes; the second part assigns a four-digit number to specific materials for identification purposes.

The UN/NA identification system utilizes the UN hazard classes and divisions list in table 2–6.

Table 2–6. UN Hazard classes and divisions

Class	Division	Placard	Hazards	Examples
Class 1—Explosives				
1	1.1	EXPLOSIVE 1.1 1	Mass explosion	Black powder TNT Dynamite Blasting caps
	1.2	EXPLOSIVE 1.2 1	Projection hazard	Detonating cords Aerial flares
	1.3	EXPLOSIVE 1.3 1	Fire hazard Possible minor blast or projection hazard	Propellant explosives Liquid-fuel rocket motors
	1.4	1.4 EXPLOSIVE 1	No significant blast hazard	Practice ammunition and signal cartridges
	1.5	1.5 BLASTING AGENT 1	Very insensitive blasting agents	Prilled (pelleted) ammonium nitrate fertilizer and blasting agents
	1.6	1.6 EXPLOSIVE 1	Extremely insensitive detonating articles	Fertilizer/fuel oil mixtures.
Class 2—Gases				
2	2.1	FLAMMABLE 2	Flammable gas	Acetylene, hydrogen, methane, and propane
	2.2	NON-FLAMMABLE GAS 2	Nonflammable gas	Carbon dioxide and anhydrous ammonia
	2.3	INHALATION HAZARD 2	Poisonous gas	Phosgene, chlorine, cyanide gas, arsine, methyl bromide
	2.4	2	Corrosive gases	Ammonia, hydrogen chloride, chlorine, methylamine

Class	Division	Placard	Hazards	Examples
		OXYGEN 2	Oxygen	
Class 3—Flammable and Combustible Liquids				
3		FLAMMABLE 3	Flammable liquid	Gasoline, kerosene, and acetone
		FLAMMABLE 3	Combustible liquid	Mineral oil and no. 6 fuel oil
Class 4—Flammable Solids				
4	4.1	FLAMMABLE SOLID 4	Flammable solid	Magnesium
	4.2	SPONTANEOUSLY COMBUSTIBLE 4	Spontaneously combustible materials	Phosphorus
	4.3	DANGEROUS WHEN WET 4	Dangerous when wet	Calcium carbide, sodium hydride, magnesium powder
Class 5—Oxidizers and Organic Peroxides				
5	5.1	OXIDIZER 5.1	Oxidizers	Calcium carbide, sodium hydride, magnesium powder
	5.2	ORGANIC PEROXIDE 5.2	Organic peroxides	Ethyl ketone peroxide and peroxide
Class 6—Toxic and Infectious Substances				
6	6.1	POISON 6	Poisonous	Arsenic
	6.2	INFECTIOUS SUBSTANCE 6	Infectious substances	Rabies, HIV, and hepatitis B
Class 7—Radioactive				

Class	Division	Placard	Hazards	Examples
7		RADIOACTIVE 7	Radioactive	Cobalt, uranium, and plutonium
Class 8 – Corrosive Materials				
8		CORROSIVE 8	Corrosive	Sulfuric acid and sodium hydroxide
Class 9 - Miscellaneous				
9		9	U.S. miscellaneous	Dangerous goods Environmental hazardous substances Dangerous wastes
		9	Canadian miscellaneous	

The four-digit UN/NA identification number is required only on tank cars, cargo tanks, and portable tanks. It may be found on the face of a standard DOT placard (fig. 2–7), or it may appear on its own panel. The DOT requires that if the UN number appears on its own panel, it must be accompanied by another placard that indicates its hazard class. When hazardous materials are being transported in bulk containers, this four-digit ID is required to be displayed on all four sides of the vehicle. Note that a single four-digit ID number is not necessarily specific to one chemical. For example, *UN 1993* is commonly displayed on many tanker trucks. This number is used when transporting diesel fuel. Placards can display three pieces of information, the four-digit UN/NA number, the hazard class number and a symbol such as a flame or skull and cross bones (fig. 2–8).

Fig. 2–7. The UN/NA ID number can be included on the DOT hazard placard or on its own.

Fig. 2–8. There are different pieces of information contained on the DOT placard.

For information pertaining to the UN/NA number responders refer to the Emergency Response Guide (ERG). Following are brief definitions and explanations to the DOT nine hazard classes.

- *Class 1—explosive materials.* Class 1 has 6 divisions and has a placard color of orange. Explosives can be detonated by heat, electrical currents or shockwaves depending on the type of explosive. Potential injuries include thermal injury due to the heat from detonation as well as mechanical injury from the blast or objects thrown by the blast. The potential for chemical injuries also exists. If an explosive material has completely detonated, it may not pose an additional risk to first responders; however, rescuers should consider the possibility that there may be additional explosive materials in the same area, or in the case of a criminal or terrorist incident, that a secondary device may be present. Etiological concerns can arise from the victims' bodily fluids, and there is always the possibility that asphyxiation could be an issue due to depleted oxygen levels in the blast area.

- *Class 2—gases.* Class 2 has three divisions and can be found on red, green, white or yellow placards. If the gas is stored under pressure, failure of the cylinder can result in mechanical, thermal, chemical, or asphyxiate hazard, depending on the material's hazard. If the contents are cryogenic in nature, a thermal hazard is present, and in confined areas asphyxiation, either simple or chemical, is a concern.

- *Class 3—flammable liquids.* Class 3 does not have divisions but is comprised of flammable and combustible liquids which is separated by the flashpoint. Placard color for this class is typically red. Substances with a flashpoint of less than 100°F are flammable and over 100°F are combustible. The risk associated with flammable liquids is that they can present a thermal and chemical hazard as well as a mechanical hazard if an explosion were to occur.

- *Class 4—flammable solids.* Class 4 has three divisions: flammable solids, dangerous when wet, and spontaneously combustible materials. Placard colors are red and white candy stripes, light blue and white on the top and red on the bottom. All can cause thermal, chemical, and mechanical hazards. As with flammable liquids they can burn hotter than the average structure fire, and

appropriate extinguishment methods for the material involved should be employed to protect the rescuer. Dangerous when wet materials will emit toxic and/or flammable gases when exposed to water. Some of them will react when exposed to high humidity. Spontaneously combustible materials will undergo combustion when exposed to air.

- *Class 5—oxidizers.* Class 5 has two divisions: oxidizers and organic peroxides. The key hazard found in oxidizers are that they enhance the burning process. This usually occurs by the compound yielding oxygen. Placard color for this class is yellow. Oxidizers and organic peroxides may react violently and without warning when contaminated or exposed to heat. Organic peroxides are also known to react explosively and are often used by terrorist when making explosives. They present a thermal, chemical, and mechanical harm concern. Due to the inherent instability of these materials, rescue operations need careful evaluation. If the incident appears to be worsening, first responders should be removed from the area.

- *Class 6—poisonous/infectious materials.* Class 6 has 2 divisions: poisons and biohazards. Placard color is white. PPE specific to the hazard must be used. These materials present a hazard by inhalation, ingestion, absorption, or injection. Certain substances may also present a thermal hazard due to their flammability. By their nature, poisonous and infections materials pose a danger to life.

- *Class 7—radioactive materials.* Class 7 has no divisions and the placard color is white and yellow. The radiological hazard is relative to the dosage rate received over time. Use the time, distance, and shielding approach to reduce harm to responders. Some radioactive materials may also present a chemical hazard. Responders may find themselves with very limited PPE and resources to deal with radioactive hazards.

- *Class 8—corrosives.* Class 8 has no divisions with a placard color of white on top and black on the bottom. Corrosive can cause full thickness destruction of human tissue and often react to metals. Corrosives may also present a chemical and potential thermal hazard. They can also present a mechanical hazard if they weaken the structure of a container or building.

- *Miscellaneous materials.* Identification of specific material(s) involved is necessary in order to determine the specific hazards to responders. Responders should be prepared to contend with any of the listed types of harm.

Global Harmonization System. In 2003, the United Nations (UN) adopted the Globally Harmonized System of Classification and Labeling of Chemicals (GHS). The full title is, *The Globally Harmonized System of Classification and Labeling of Chemicals.* The GHS is a system for standardizing and harmonizing the classification and labeling of chemicals on a worldwide basis. The GHS includes criteria for the classification of health, physical, and environmental hazards, as well as the primary goal of standardizing and harmonizing the labeling of chemicals.

The main strategic issues behind the GHS are:

- Defining health, physical and environmental hazards of chemicals

- Creating classification processes that use available data on chemicals for comparison with the defined hazard criteria

- Communicating hazard information, as well as protective measures, on labels and Safety Data Sheets (SDS)

It should be noted that prior to OSHA adopting and enforcing the GHS in 2017, chemicals and their properties were listed on Materials Safety Data sheets (MSDS). The GHS has re-termed these documents by simply calling them Safety Data Sheets or SDS.

The GHS has been adopted by OHSA and is now an OSHA standard in the United States. The GHS Document (referred to as "The Purple Book," as shown in fig. 2–9) establishes hazard classification and communication provisions with explanatory information on how to apply the system. The elements in the GHS have created tools to meet the basic requirements of any hazard communication system, which determines if the chemical product produced and/or supplied is hazardous and how to prepare a label and/or Safety Data Sheet. Regulatory authorities in countries adopting the GHS take the agreed criteria and provisions, and implement them through their own regulatory process and procedures rather than simply incorporating the text of the GHS into their national requirements.

Fig. 2–9. The Purple Book

The main elements of the hazard classification criteria are summarized below:

Physical hazards. Substances or articles are assigned to nine different hazard classes largely based on the United Nations Dangerous Goods System.

1. Explosives, which are assigned to one of six subcategories depending on the type of hazard they present, as used in the UN Dangerous Goods System.

2. Gases are category 1 flammable if they start to flame in a range in air at 68°F (20°C) and a standard pressure of 101.3 kPa. Category 2 is Non flammable and non toxic gases, and category 3 is toxic gases. Substances and mixtures of this hazard class are assigned to one of two hazard categories on the basis of the outcome of the test or calculation method.

3. A flammable liquid is a liquid with a flash point of not more than 199°F (93°C). Substances and mixtures of this hazard class are assigned to one of four hazard categories on the basis of the flash point and boiling point. A pyrophoric liquid is a liquid that, even in small quantities, is liable to ignite within five minutes after coming into contact with air.

4. A flammable solid is one that is readily combustible or may cause or contribute to fire through friction. Readily combustible solids are powdered, granular, or pasty substances which are dangerous if they can

be easily ignited by brief contact with an ignition source, such as a burning match, and if the flame spreads rapidly. It is further divided into

- Flammable solids,

- Polymerizing substances, and

- Self-reactive substances, i.e., thermally unstable solids liable to undergo a strongly exothermic thermal decomposition even without participation of oxygen (air) other than materials classified as explosive, organic peroxides or as oxidizing.

  Spontaneously combusting solids or pyrophoric solids are solids that, even in small quantities, are liable to ignite within five minutes after coming into contact with air. Substances and mixtures of this hazard class are assigned to a single hazard category on the basis of the outcome of the UN Test N.2. Self-heating substances are solids or liquids, other than a pyrophoric substance, which, by reaction with air and without energy supply, are liable to self-heat. Substances and mixtures of this hazard class are assigned to one of two hazard categories on the basis of the outcome of the UN Test N.4. Substances which on contact with water emit flammable gases are liable to become spontaneously flammable or to give off flammable gases in dangerous quantities. Substances and mixtures of this hazard class are assigned to one of three hazard categories on the basis of the outcome of UN Test N.5, which measures gas evolution and speed of evolution. Flammable aerosols can be classified as Class 1 or Class 2 if they contain any component, which is classified as flammable.

5. Oxidizing substances and organic peroxides contain

  - Category 1: oxidizing substances and

  - Category 2: organic peroxides, organic liquids or solids that contain the bivalent -O-O- structure and may be considered a derivative of hydrogen peroxide, where one or both of the hydrogen atoms have been replaced by organic radicals. The term also includes organic peroxide formulations (mixtures).

Substances and mixtures of this hazard class are assigned to one of seven 'Types', A to G, on the basis of the outcome of the UN Test Series A to H.

6. Toxic and infectious substances

7. Radioactive substances

8. Substances corrosive to metal are substances or mixtures that by chemical action will materially damage or even destroy metals. These substances or mixtures are classified in a single hazard category on the basis of tests (Steel: ISO 9328 (II): 1991 - Steel type P235; Aluminum: ASTM G31-72 (1990) - non-clad types 7075-T6 or AZ5GU-T66). The GHS criteria are a corrosion rate on steel or aluminum surfaces exceeding 6.25 mm per year at a test temperature of 55°C.

9. Miscell**aneous** dangerous substances which includes environmentally dangerous substances

Health hazards

- **Acute toxicity** includes five GHS categories from which the appropriate elements relevant to transport, consumer, worker and environment protection can be selected. Substances are assigned to one of the five **toxicity** categories on the basis of LD_{50} (oral, dermal) or LC_{50} (inhalation).

- **Skin corrosion** means the production of irreversible damage to the skin following the application of a test substance for up to 4 hours. Substances and mixtures in this hazard class are assigned to a single harmonized corrosion category.

- **Skin irritation** means the production of reversible damage to the skin following the application of a test substance for up to 4 hours. Substances and mixtures in this hazard class are assigned to a single **irritant** category. For those authorities, such as **pesticide** regulators, wanting more than one designation for skin irritation, an additional mild irritant category is provided.

- **Serious eye damage** means the production of tissue damage in the eye, or serious physical decay of vision, following application of a test substance to the front surface of the eye, which is not fully reversible within 21 days of application. Substances and mixtures in this hazard class are assigned to a single harmonized category.

- **Eye irritation** means changes in the eye following the application of a test substance to the front surface of the eye, which are fully reversible within 21 days of application. Substances and mixtures in this hazard class are assigned to a single harmonized hazard category. For authorities, such as pesticide regulators, wanting more than one designation for eye irritation, one of two subcategories can be selected, depending on whether the effects are reversible in 21 or 7 days.

- **Respiratory sensitizer** means a substance that induces **hypersensitivity** of the airways following inhalation of the substance. Substances and mixtures in this hazard class are assigned to one hazard category.

- **Skin sensitizer** means a substance that will induce an allergic response following skin contact. The definition for "skin **sensitizer**" is equivalent to "contact sensitizer." Substances and mixtures in this hazard class are assigned to one hazard category.

- **Germ cell mutagenicity** means an agent giving rise to an increased occurrence of **mutations** in populations of cells and/or organisms. Substances and mixtures in this hazard class are assigned to one of two hazard categories. Category 1 has two subcategories.

- **Carcinogenicity** means a chemical substance or a mixture of chemical substances that induce **cancer** or increase its incidence. Substances and mixtures in this hazard class are assigned to one of two hazard categories. Category 1 has two subcategories.

- **Reproductive toxicity** includes adverse effects on sexual function and **fertility** in adult males and females, as well as **developmental toxicity** in offspring. Substances and mixtures with reproductive and/or developmental effects are assigned to one of two hazard categories, 'known or presumed' and 'suspected'. Category 1 has two subcategories for reproductive and developmental effects. Materials which cause concern for the health of breastfed children have a separate category: effects on or via **Lactation**.

- **Specific target organ toxicity (STOT)** [4] category distinguishes between single and repeated exposure for Target Organ Effects. All significant health effects, not otherwise specifically included in the GHS, that can impair function, both reversible and irreversible, immediate and/or delayed are included in the non-lethal target organ/systemic toxicity class (TOST). Narcotic effects and respiratory tract irritation are considered to be target organ systemic effects following a single exposure. Substances and

mixtures of the single exposure target organ toxicity hazard class are assigned to one of three hazard categories. Substances and mixtures of the repeated exposure target organ toxicity hazard class are assigned to one of two hazard categories.

- **Aspiration hazard** includes severe acute effects such as **chemical pneumonia**, varying degrees of **pulmonary injury** or death following **aspiration**. Aspiration is the entry of a liquid or solid directly through the oral or nasal cavity, or indirectly from vomiting, into the **trachea** and lower respiratory system. Substances and mixtures of this hazard class are assigned to one of two hazard categories this hazard class on the basis of **viscosity**.

Environmental hazards. **Acute aquatic toxicity** means the intrinsic property of a material of causing injury to an aquatic organism in a short-term exposure. Substances and mixtures of this hazard class are assigned to one of three toxicity categories on the basis of acute toxicity data: LC_{50} (**fish**) or EC_{50} (**crustacean**) or ErC_{50} (for **algae** or other **aquatic plants**). In some regulatory systems these acute toxicity categories may be subdivided or extended for certain sectors.

Chronic aquatic toxicity means the potential or actual properties of a material to cause adverse effects to aquatic organisms during exposures that are determined in relation to the lifecycle of the organism. Substances and mixtures in this hazard class are assigned to one of four toxicity categories on the basis of acute data and environmental fate data: LC_{50} (fish) or EC_{50} (crustacea) or ErC_{50} (for algae or other aquatic plants) and **degradation** or **bioaccumulation**.

Classification of mixtures. The GHS approach to the classification of mixtures for health and environmental hazards is also complex. The classification of mixtures is based on the following steps:

1. Where **toxicological** or ecotoxicological test data are available for the mixture itself, the classification of the mixture will be based on that data;

2. Where test data are not available for the mixture itself, then the appropriate bridging principles should be applied, which uses test data for components and/or similar mixtures;

3. If (1) test data are not available for the mixture itself, and (2) the bridging principles cannot be applied, then use the calculation or cutoff values described in the specific endpoint to classify the mixture.

The standardized label elements included in the GHS are:

- **Symbols (GHS hazard pictograms)**: Convey **health**, physical and **environmental** hazard information, assigned to a GHS hazard class and category. Pictograms include the harmonized hazard symbols plus other graphic elements, such as borders, background patterns or colors and substances which have target organ toxicity. Also, harmful chemicals and irritants are marked with an **exclamation mark**, replacing the European **saltire**. Pictograms will have a black symbol on a white background with a red diamond frame. For transport, pictograms will have the background, symbol and colors currently used in the **UN Recommendations on the Transport of Dangerous Goods**. Where a transport pictogram appears, the GHS pictogram for the same hazard should not appear.

- **Signal words**: "Danger" or "Warning" will be used to emphasize hazards and indicate the relative level of severity of the hazard, assigned to a GHS hazard class and category. Some lower level hazard categories do not use signal words. Only one signal word corresponding to the class of the most severe hazard should be used on a label.

- **GHS hazard statement**: Standard phrases assigned to a **hazard class** and category that describe the nature of the hazard. An appropriate statement for each GHS hazard should be included on the label for products possessing more than one hazard.

The additional label elements included in the GHS are:

- **GHS precautionary statements**: Measures to minimize or prevent **adverse effects**. There are four types of precautionary statements covering: prevention, **response in cases of accidental spillage or exposure**, storage, and disposal. The precautionary statements have been linked to each GHS hazard statement and type of hazard.

- **Product identifier** (ingredient disclosure): Name or number used for a hazardous product on a label or in the SDS. The GHS label for a substance should include the chemical identity of the substance. For mixtures, the label should include the chemical identities of all ingredients that contribute to acute toxicity, skin corrosion or serious eye damage, germ cell mutagenicity, carcinogenicity, reproductive toxicity, skin or respiratory sensitization, or Systemic Target Organ Systemic (STOT), when these hazards appear on the label.

- **Supplier identification**: The name, address and telephone number should be provided on the label.

- **Supplemental information**: Non-harmonized information on the container of a hazardous product that is not required or specified under the GHS. Supplemental information may be used to provide further detail that does not contradict or cast doubt on the validity of the standardized hazard information.

GHS label format. The GHS includes directions for application of the hazard communication elements on the label. It specifies for each hazard, and for each class within the hazard, what signal word, **pictogram**, and hazard statement should be used. The GHS hazard pictograms, signal words and hazard statements should be located together on the label. Transport pictograms are different in appearance than the GHS pictograms. Annex 7 of the Purple Book explains how the GHS pictograms are expected to be proportional to the size of the label text so that generally the GHS pictograms would be smaller than the transport pictograms.

Safety data sheet. The **safety data sheet** or SDS (The GHS dropped the word "material" from material safety data sheet) is specifically aimed at use in the workplace however emergency responders use SDS's exclusively during emergency incidents. An SDS should provide comprehensive information about the chemical product that allows employers and workers to obtain concise, relevant and accurate information in perspective to the hazards, uses and **risk management** of the chemical product in the workplace. The following are the 16 sections of an SDS:

1. Identification
2. Hazard(s) identification
3. Composition/ information on ingredients
4. **First-aid** measures
5. Firefighting measures
6. **Accidental release measures**
7. Handling and storage
8. Exposure control/ personal protection
9. Physical and chemical properties

10. **Chemical stability** and **reactivity**

11. Toxicological information

12. Ecological information

13. Disposal considerations

14. Transport information

15. Regulatory information

16. Other information

Beware of Conflicting Information—NFPA 704 vs. GHS. Relevant to NFPA Standard 704 and its placard system for fixed vessels in fixed facilities, the GHS system is in conflict with the color cart and the numbering of hazard levels. There is also conflict with the HMIS which mimics the NFPA 704 system. While the colors remain the same (red = flammability, blue = health, yellow = reactivity, and white for special instructions) the numbering system for the levels of hazard are opposite of each other. Note the chart below.

NFPA 704 and HMIS Hazard Ratings	GHS Hazard Categories
0 = Minimal or no hazard	Category 1 = Severe hazard
1 = Slight hazard	Category 2 = Serious hazard
2 = Moderate hazard	Category 3 = Moderate hazard
3 = Serious hazard	Category 4 = Slight hazard
4 = Severe hazard	Category 5 = Minimal hazard

Special situation transportation markings. The marine pollutant marking is displayed on bulk shipping packages containing materials designated as marine pollutants. This marking must be displayed on both ends and both sides of the container unless it has been properly placarded as corrosive, flammable gas, etc. (fig. 2–10).

Fig. 2–10. Marine pollutant marking

The **elevated temperature materials** marking is displayed on each side and each end of bulk containers with the word HOT. The word HOT can be placed directly on the container or written in black letters and placed on a placard with a white background (fig. 2–11).

Fig. 2–11. Elevated temperature marking

Elevated-temperature materials are those materials that, when transported in bulk containers, are:

- Liquids at or above 212°F (100°C)

- Liquids with a flash point at or above 100°F (37.8°C) that are intentionally heated and then transported at temperatures at or above their flash point

- Solids at a temperature at or above 464°F (240°C)

All bulk containers containing materials that meet these requirements must be marked with the word HOT unless they contain molten aluminum or molten sulfur. These containers must be marked MOLTEN ALUMINUM and MOLTEN SULFUR, respectively.

The **inhalation hazard** marking is used on materials considered to be hazardous when inhaled (fig. 2–12). An example is anhydrous ammonia.

INHALATION HAZARD

Fig. 2–12. Inhalation hazard marking

Pipeline markings. A large network of pipelines runs throughout the United States, and millions of gallons of hazardous materials are transported through it each year. Like other transport methods, these pipelines fall under the jurisdiction of the DOT. Although primarily underground, they can pass over or under roads, railroads, and waterways. Typically, gas pipelines are used for one specific product only. Liquid petroleum pipelines, however, may carry several different petroleum products at the same time. It is imperative to recognize pipeline markings inside and adjacent to fixed facilities and throughout your response district. Pipeline markings may not always mark the exact location of the pipeline, and they are typically found at intersections with roads and railroads. They are required to display the following information (fig. 2–13):

- Contents of the pipeline (product class)

- Signal word: caution, warning, danger

- Operator of the pipeline

- Emergency contact telephone number

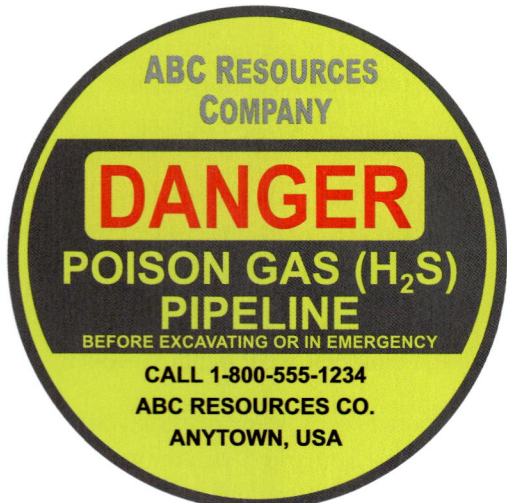

Fig. 2–13. Pipeline markings are required to contain several pieces of information namely the owner's name and an emergency phone number.

Military markings. Military markings are used to identify detonation, fire, and special hazards on military transport vehicles and at U.S. military installations.

Because military ordinance is designed to create bodily harm and property damage, extreme caution should be practiced upon approach. Responders should remember that drivers may be under orders not to identify cargo that is sensitive in nature. The military should always be contacted immediately if one of their transport vehicles is involved in an incident, as they can provide valuable assistance in handling the product. Tables 2–7, 2–8, and 2–9 illustrate the symbols used by the military for hazardous materials as well as a general description of their properties.

Table 2–7. Military explosive/fire symbols

Symbol	Hazard
1	Mass detonation hazard
2	Explosive with fragmentation hazard
3	Mass fire hazard
4	Moderate fire hazard

Table 2–8. Military chemical hazard symbols

Symbol	Hazard
	Highly toxic
	Harassing agents
	White phosphorus munitions
	Apply no water
	Wear protective breathing apparatus

Table 2–9. Military chemical warfare agents

Symbol	Hazard
G	G-type nerve agents
H	H-type mustard agents
L	Lewisite
VX	VX nerve agents

Radioactive labels. Radioactive labels contain several key points of information about the material being shipped and the level of danger associated with it. Responders need to be able to recognize and understand the importance of this information when coming in contact with a radioactive material that is being shipped. Figure 2–14 illustrates the different types of information presented on a radioactive label.

Fig. 2–14. Radioactive package labels contain key information about the radioactivity of the material being shipped.

The radioactive label identifies the following:

- The *visual hazard* symbol indicates a radioactive material as does the *UN class number* at the bottom.

- The *vertical bars* indicate the shipping category as either radioactive I, II, or III.

- The *contents* line identifies the material being shipped in the package.

- The *activity* line indicates the level of radiation emitted by the package.

- The *transport index* is the maximum level of radiation at a point 1 meter away from the surface of the package.

- There are three different shipping categories for radioactive materials. These are based on the level of radiation emitted by the material and the hazards associated with it. Additionally, fissile labels are used on packages containing fissile material. The fissile label will be displayed with one of the three radioactive category labels (table 2–10). The criticality safety index on this label is used by the shipper to indicate the maximum number of packages or containers allowed in a single shipment.

Table 2–10. Radioactive material and fissile labels

Category	Symbol	Color	Definition
I	RADIOACTIVE I	White	Almost no radiation. Maximum allowable radioactivity on the package surface is 0.5 mrem/hr.
II	RADIOACTIVE II	Yellow/White	Low radiation levels. Maximum level of radioactivity of the package surface is 50 mrem/hr and 10 mrem/hr at a distance of 3 ft (1 m) from the package.
III	RADIOACTIVE III	Yellow/White	Higher radiation levels. Maximum allowable radioactivity of the package surface is 200 mrem/hr and 10 mrem/hr at a distance of 3 ft (1 m) from the package. Vehicles carrying Yellow III packages must be placarded on both sides and ends.
Fissile	FISSILE		Fissile material label will be added to category I, II, or III labels.

Miscellaneous container markings—transportation.

In addition to the placards and labels used in transportation discussed previously, there are other markings that provide helpful information. Identification numbers on the container can be compared to the shipping papers to ensure that they match; they also allow the responder to contact the shipper directly and confirm the material being shipped against the tank ID number.

On highway cargo tanks, in addition to the standard DOT placards, the shipper's ID or trailer number may be found (fig. 2–15). Other markings may include the company name and logo, a specially colored tank designated for the material being transported, and the manufacturer's specification plate.

Fig. 2–15. Shipper's ID or trailer number can aid in the identification process.

On rail transport cars, the name of the material or commodity may be stenciled directly onto the tank for specific tank cars. These dedicated rail cars will have the product name stenciled on the side and are only used to transport that particular material. Reporting marks and volume/specifications markings can also be found (fig. 2–16). The seven-digit standard transportation commodity code (STCC) may be stenciled on the tank. Numbers starting with 48 indicate a hazardous waste.

Fig. 2–16. Reporting marks or volume specifications may be stenciled on the tank.

Lastly, on intermodals, additional markings can be found on the tank or tank frame. These markings indicate the ownership of the tank, the standards to which it was built, and a data plate indicating the size, type, and country codes (fig. 2–17).

Fig. 2–17. Typical intermodal container markings

Miscellaneous markings

Pesticide labels fall under the category of miscellaneous markings. They are used on containers that store chemicals designed to kill insects; these can be found on materials in fixed facilities or in transport. Pesticides can be found in all types of occupancies ranging from commercial to residential and in varying quantity, depending on their use. These labels contain the name of the pesticide and the following information (fig. 2–18):

- A signal word is used to indicate how toxic the materials are. They are broken into the following categories:
  - *Danger/poison* means highly toxic.
  - *Warning* means moderately toxic.
  - *Caution* means relatively low toxicity.

The label may also contain a *precautionary statement*, such as *keep from waterways* or *keep out of reach of children*.

- Hazard statements list special flammability, explosion, or chemical hazards as well as if it is an environmental hazard. For example: Extremely Flammable—flash point below 80°F (27°C).

- The active ingredient section lists the names and percentage of the active chemicals that make up the pesticide. Inert ingredients are listed as a percentage only.

- Finally, the *Environmental Protection Agency (EPA)* registration number can be found on the label. The

EPA regulates the manufacturing and labeling of pesticides and requires a registration number on the label. Additional information can be obtained by contacting the Chemical Transportation Emergency Center (CHEMTREC) and providing it with this number. The Canadian version of this number is the pest control products (PCP) number.

The label may also contain information on the mode of entry into the body, storage and disposal requirements, first aid information, and, if known, an antidote in the event of poisoning.

Fig. 2–18. Pesticide labels contain the name of the material and hazard level.

PRINTED RESOURCES

U.S. DOT Emergency Response Guidebook

The **U.S. Department of Transportation Emergency Response Guidebook** (commonly referred to as the **ERG**) is the most common and widely used resource by emergency responders at hazardous materials/ WMD incidents (fig. 2–19). It outlines the general concerns of the product as well as recommended initial actions to take, and is the first source of information used by responders in hazardous materials transportation incidents. The ERG is intended to give the first responder quick information regarding the hazards and protective actions in a hazardous materials incident. It is meant as an initial reference guide, but its usefulness is limited to the first few minutes of the incident.

Other resources need to be referenced for more specific product information.

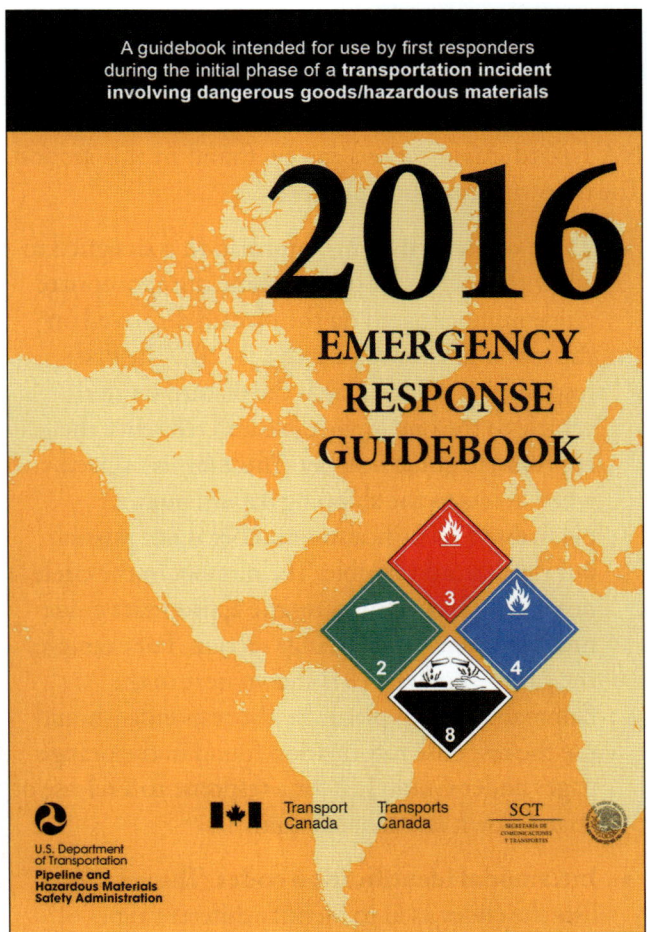

Fig. 2–19. Department of Transportation (DOT) Emergency Response Guidebook (ERG)

The ERG is released every four years. Copies are provided by the U.S. Department of Transportation to the State Emergency Management Agency, and are then distributed to the local responders for use in each vehicle assigned to those agencies. *Every fire department response vehicle should have a U.S. DOT Emergency Response Guidebook.* If you do not have one in your apparatus, ask your senior officer to have one placed in the vehicle. The ERG is available for download from various websites, and can be found by a web search for "DOT Emergency Response Guidebook."

The ERG is organized into five distinct color-coded sections designed to make the guide simple to use. The white section contains important reference material, the yellow section can be used to look up a material by its four-digit UN/NA ID number, the blue section references the material by its name, the orange section contains the emergency response guides, and the green section provides the initial isolation and protective action distances for toxic by inhalation materials. The following covers each section in greater detail.

White section. Found at the beginning of the book, the white section provides valuable information on how to use the ERG as well as how the information is compiled. A list of important phone numbers as well as the UN/NA hazard classifications can be found in this section. The following can also be found:

- **Table of placards.** This information is designed to allow the responder to determine the appropriate guide number based on the placard displayed on the road trailer or rail car (fig. 2–20). The first responder is able to match the placard on the vehicle to the one in the table and reference the appropriate guide number. This table is utilized when access to the shipping papers, number placards, or identification number is not possible, but the placard is visible. The number in the circle near each placard indicates the appropriate guide number found in the orange section. For example, the guide number 118 is shown next to the flammable liquids placards. The responder should use the response information found in the orange pages under Guide 118 for a response to a release of an unknown flammable liquid.

- **Intermodal identification codes.** This section lists the codes used to identify different types of materials shipped in intermodal containers.

- **Pipeline transportation.** This section reviews the markings used to indicate the presence of a pipeline as well as indications of a potential leak.

The back of the book contains another section, color coded white, that contains general information about personal protective equipment, fire and spill control procedures, responses involving terrorist incidents, as well as a glossary and additional contact phone numbers.

- **Railcar and road trailer guide.** In the absence of a placard or other identifying marks, this reference can be used to determine the appropriate guide based on the container type (fig. 2–21). The number in the circle near each placard indicates the appropriate guide number. Because container designs vary greatly, the suggested guides should be used only as a last resort if no other means of identification is available.

Yellow section—identification by UN/NA ID number. The yellow-bordered pages of the ERG list the UN/NA ID numbers in numerical order and should be used when the product ID number is known (fig. 2–22). The ID number is the four-digit number displayed on all four sides of a vehicle or container transporting hazardous materials. This section provides the shipping name and the guide number assigned to the product. Highlighted chemicals indicate a poison or inhalation poison risk and require referencing the Initial Isolation and Protective Action Distances information found in the green section. This is addressed in greater detail shortly.

Blue section—identification by name. The blue section lists material names in alphabetical order and should be used when the product's name is known (fig. 2–23). The names of materials in this section are listed as they appear on shipping documents. The responder should carefully double check the spelling. This section provides the UN/NA ID number and the guide number assigned to the product. Highlighted chemicals indicate a poison or inhalation poison risk and require referencing the Initial Isolation and Protective Action Distances information found in the green section. This is addressed in greater detail shortly.

Orange section—initial action guides. The orange section identifies the most significant hazards and informs first responders of initial actions to be taken to protect from immediate hazards. Each guide is broken down into three sections. The first section lists the *potential hazards*, listing the product's health effects and fire/explosion hazard. The hazard listed first represents the more serious danger. The next section outlines the *public safety measures* that should be taken in the event of a release and addresses isolation distances as well as personal protective clothing and respiratory recommendations. The protective clothing options include street clothing and work uniforms, structural firefighting protective clothing, positive-pressure SCBA, and chemical protective clothing and equipment. These protective clothing options are addressed in greater detail in chapter 3. Finally, the emergency response section covers actions to be taken in the event of a fire, spill, or leak, as well as first-aid information. Each guide is two pages with health and safety information included on the left page (fig. 2–24) and emergency response information on the right page (fig. 2–25). Each guide covers a group of chemicals with similar properties.

RESPONSE GUIDE TO USE ON-SCENE
USING THE SHIPPING DOCUMENT, NUMBERED PLACARD, OR ORANGE PANEL NUMBER

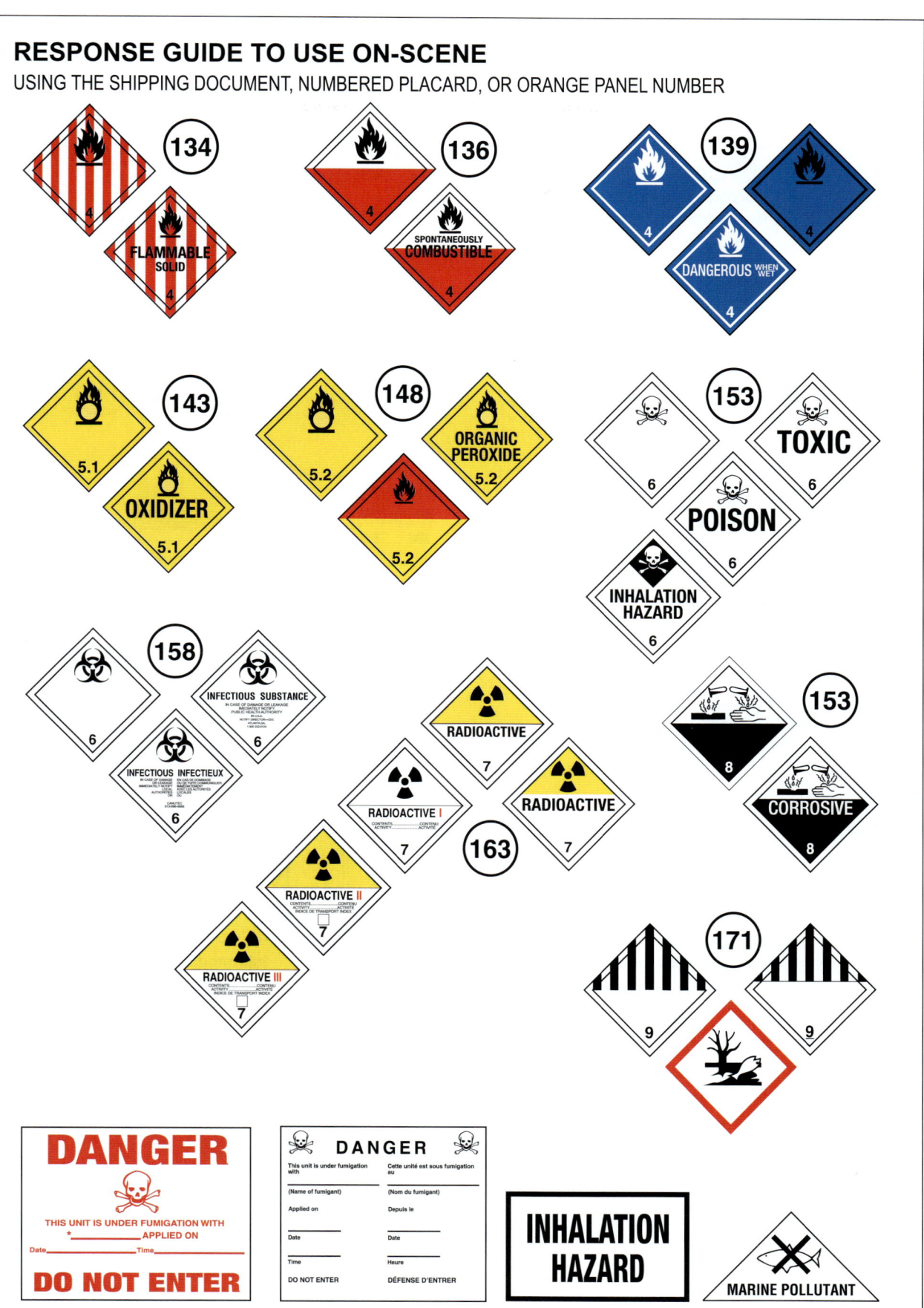

Fig. 2—20. ERG—Table of placards

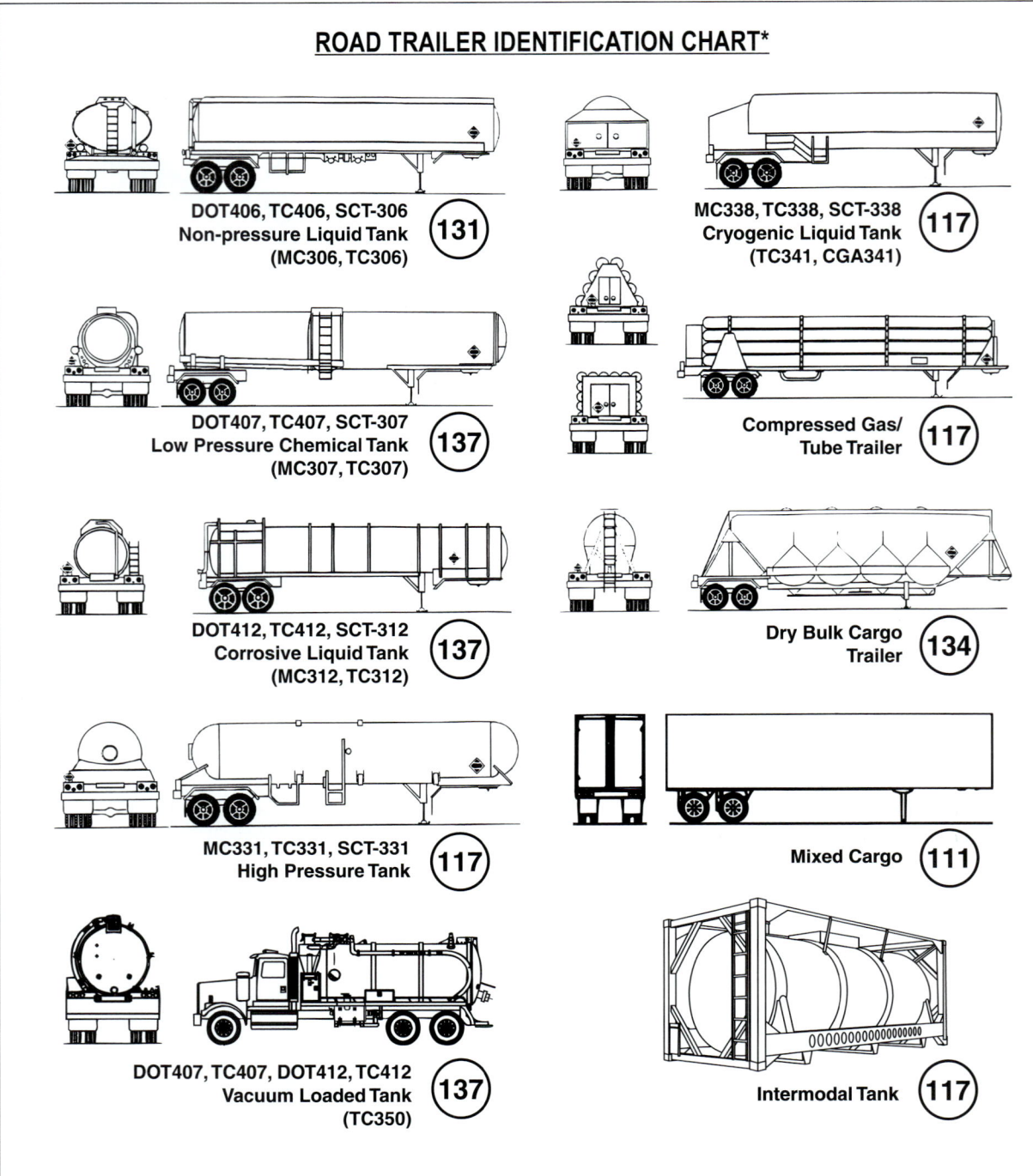

CAUTION: This chart depicts only the most general shapes of road trailers. Emergency response personnel must be aware that there are many variations of road trailers, not illustrated above, that are used for shipping chemical products. The suggested guides are for the most hazardous products that may be transported in these trailer types.

*   **The recommended guides should be considered as a last resort if the material cannot be identified by any other means.**

Fig. 2–21. ERG—Railcar and road trailer guide

ID No.	Guide No.	Name of Material
1493	140	Silver nitrate
1494	141	Sodium bromate
1495	140	Sodium chlorate
1496	143	Sodium chlorite
1498	140	Sodium nitrate
1499	140	Potassium nitrate and Sodium nitrate mixture
1499	140	Sodium nitrate and Potassium nitrate mixture
1500	140	Sodium nitrite
1502	140	Sodium perchlorate
1503	140	Sodium permanganate
1504	144	Sodium peroxide
1505	140	Sodium persulfate
1505	140	Sodium persulphate
1506	143	Strontium chlorate
1506	143	Strontium chlorate, solid
1506	143	Strontium chlorate, solution
1507	140	Strontium nitrate
1508	140	Strontium perchlorate
1509	143	Strontium peroxide
1510	143	Tetranitromethane
1511	140	Urea hydrogen peroxide
1512	140	Zinc ammonium nitrite
1513	140	Zinc chlorate
1514	140	Zinc nitrate
1515	140	Zinc permanganate
1516	143	Zinc peroxide
1517	113	Zirconium picramate, wetted with not less than 20% water
1541	155	Acetone cyanohydrin, stabilized
1544	151	Alkaloids, solid, n.o.s. (poisonous)

ID No.	Guide No.	Name of Material
1544	151	Alkaloid salts, solid, n.o.s. (poisonous)
1545	155	Allyl isothiocyanate, stabilized
1546	151	Ammonium arsenate
1547	153	Aniline
1548	153	Aniline hydrochloride
1549	157	Antimony compound, inorganic, n.o.s.
1549	157	Antimony compound, inorganic, solid, n.o.s.
1549	157	Antimony tribromide, solid
1549	157	Antimony tribromide, solution
1549	157	Antimony trifluoride, solid
1549	157	Antimony trifluoride, solution
1550	151	Antimony lactate
1551	151	Antimony potassium tartrate
1553	154	Arsenic acid, liquid
1554	154	Arsenic acid, solid
1555	151	Arsenic bromide
1556	152	Arsenic compound, liquid, n.o.s.
1556	152	Arsenic compound, liquid, n.o.s., inorganic
1556	152	MD
1556	152	Methyldichloroarsine
1556	152	PD
1557	152	Arsenic compound, solid, n.o.s.
1557	152	Arsenic compound, solid, n.o.s., inorganic
1557	152	Arsenic sulfide
1557	152	Arsenic sulphide
1557	152	Arsenic trisulfide
1557	152	Arsenic trisulphide
1558	152	Arsenic
1559	151	Arsenic pentoxide

Fig. 2–22. ERG—identification by UN/NA ID number

Name of Material	Guide No.	ID No.	Name of Material	Guide No.	ID No.
Chlorodinitrobenzenes, solid	153	3441	Chlorophenates, solid	154	2905
1-Chloro-2,3-epoxypropane	131P	2023	Chlorophenolates, liquid	154	2904
2-Chloroethanal	153	2232	Chlorophenolates, solid	154	2905
Chloroform	151	1888	Chlorophenols, liquid	153	2021
Chloroformates, n.o.s.	155	2742	Chlorophenols, solid	153	2020
Chloroformates, poisonous, corrosive, flammable, n.o.s.	155	2742	Chlorophenyltrichlorosilane	156	1753
			Chloropicrin	154	1580
Chloroformates, poisonous, corrosive, n.o.s.	154	3277	Chloropicrin and Methyl bromide mixture	123	1581
Chloroformates, toxic, corrosive, flammable, n.o.s.	155	2742	Chloropicrin and Methyl chloride mixture	119	1582
Chloroformates, toxic, corrosive, n.o.s.	154	3277	Chloropicrin mixture, n.o.s.	154	1583
			Chloropivaloyl chloride	156	9263
Chloromethyl chloroformate	157	2745	Chloroplatinic acid, solid	154	2507
Chloromethyl ethyl ether	131	2354	Chloroprene, stabilized	131P	1991
3-Chloro-4-methylphenyl isocyanate	156	2236	1-Chloropropane	129	1278
3-Chloro-4-methylphenyl isocyanate, liquid	156	2236	2-Chloropropane	129	2356
			3-Chloropropanol-1	153	2849
3-Chloro-4-methylphenyl isocyanate, solid	156	3428	2-Chloropropene	130P	2456
			2-Chloropropionic acid	153	2511
Chloronitroanilines	153	2237	2-Chloropropionic acid, solid	153	2511
Chloronitrobenzenes	152	1578	2-Chloropropionic acid, solution	153	2511
Chloronitrobenzenes, liquid	152	1578	2-Chloropyridine	153	2822
Chloronitrobenzenes, liquid	152	3409	Chlorosilanes, corrosive, flammable, n.o.s.	155	2986
Chloronitrobenzenes, solid	152	1578	Chlorosilanes, corrosive, n.o.s.	156	2987
Chloronitrotoluenes	152	2433	Chlorosilanes, flammable, corrosive, n.o.s.	155	2985
Chloronitrotoluenes, liquid	152	2433			
Chloronitrotoluenes, solid	152	2433	Chlorosilanes, n.o.s.	155	2985
Chloronitrotoluenes, solid	152	3457	Chlorosilanes, n.o.s.	155	2986
Chloropentafluoroethane	126	1020	Chlorosilanes, n.o.s.	156	2987
Chloropentafluoroethane and Chlorodifluoromethane mixture	126	1973	Chlorosilanes, n.o.s.	139	2988
			Chlorosilanes, poisonous, corrosive, flammable, n.o.s.	155	3362
Chlorophenates, liquid	154	2904			

Fig. 2–23. ERG—Identification by material name

GUIDE 116 GASES - FLAMMABLE (UNSTABLE)

POTENTIAL HAZARDS

FIRE OR EXPLOSION

- **EXTREMELY FLAMMABLE.**
- Will be easily ignited by heat, sparks or flames.
- Will form explosive mixtures with air.
- Silane (UN2203) will ignite spontaneously in air.
- Those substances designated with a **(P)** may polymerize explosively when heated or involved in a fire.
- Vapors from liquefied gas are initially heavier than air and spread along ground.
- Vapors may travel to source of ignition and flash back.
- Cylinders exposed to fire may vent and release flammable gas through pressure relief devices.
- Containers may explode when heated.
- Ruptured cylinders may rocket.

HEALTH

- Vapors may cause dizziness or asphyxiation without warning.
- Some may be toxic if inhaled at high concentrations.
- Contact with gas or liquefied gas may cause burns, severe injury and/or frostbite.
- Fire may produce irritating and/or toxic gases.

PUBLIC SAFETY

- **CALL EMERGENCY RESPONSE Telephone Number on Shipping Paper first. If Shipping Paper not available or no answer, refer to appropriate telephone number listed on the inside back cover.**
- As an immediate precautionary measure, isolate spill or leak area for at least 100 meters (330 feet) in all directions.
- Keep unauthorized personnel away.
- Stay upwind, uphill and/or upstream.
- Many gases are heavier than air and will spread along ground and collect in low or confined areas (sewers, basements, tanks).

PROTECTIVE CLOTHING

- Wear positive pressure self-contained breathing apparatus (SCBA).
- Structural firefighters' protective clothing will only provide limited protection.

EVACUATION

Large Spill
- Consider initial downwind evacuation for at least 800 meters (1/2 mile).

Fire
- If tank, rail car or tank truck is involved in a fire, ISOLATE for 1600 meters (1 mile) in all directions; also, consider initial evacuation for 1600 meters (1 mile) in all directions.

 In Canada, an Emergency Response Assistance Plan (ERAP) may be required for this product. Please consult the shipping document and/or the ERAP Program Section (page 391).

ERG 2016

Fig. 2–24. ERG—Response guide outlining health and safety information

Awareness Response to Hazardous Materials

HAZMAT

Chapter 2

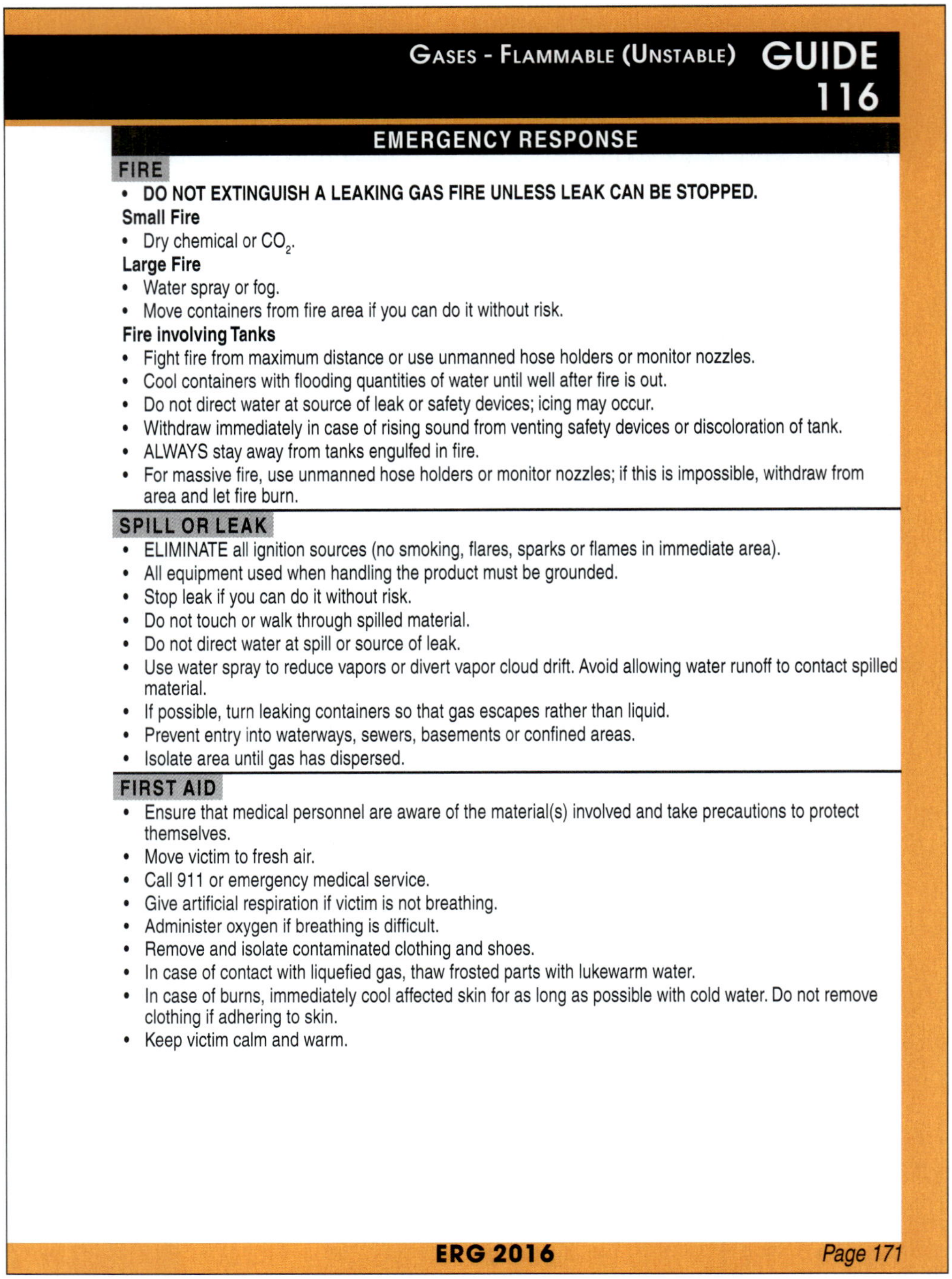

GASES - FLAMMABLE (UNSTABLE)   GUIDE 116

EMERGENCY RESPONSE

FIRE

- **DO NOT EXTINGUISH A LEAKING GAS FIRE UNLESS LEAK CAN BE STOPPED.**

Small Fire

- Dry chemical or CO_2.

Large Fire

- Water spray or fog.
- Move containers from fire area if you can do it without risk.

Fire involving Tanks

- Fight fire from maximum distance or use unmanned hose holders or monitor nozzles.
- Cool containers with flooding quantities of water until well after fire is out.
- Do not direct water at source of leak or safety devices; icing may occur.
- Withdraw immediately in case of rising sound from venting safety devices or discoloration of tank.
- ALWAYS stay away from tanks engulfed in fire.
- For massive fire, use unmanned hose holders or monitor nozzles; if this is impossible, withdraw from area and let fire burn.

SPILL OR LEAK

- ELIMINATE all ignition sources (no smoking, flares, sparks or flames in immediate area).
- All equipment used when handling the product must be grounded.
- Stop leak if you can do it without risk.
- Do not touch or walk through spilled material.
- Do not direct water at spill or source of leak.
- Use water spray to reduce vapors or divert vapor cloud drift. Avoid allowing water runoff to contact spilled material.
- If possible, turn leaking containers so that gas escapes rather than liquid.
- Prevent entry into waterways, sewers, basements or confined areas.
- Isolate area until gas has dispersed.

FIRST AID

- Ensure that medical personnel are aware of the material(s) involved and take precautions to protect themselves.
- Move victim to fresh air.
- Call 911 or emergency medical service.
- Give artificial respiration if victim is not breathing.
- Administer oxygen if breathing is difficult.
- Remove and isolate contaminated clothing and shoes.
- In case of contact with liquefied gas, thaw frosted parts with lukewarm water.
- In case of burns, immediately cool affected skin for as long as possible with cold water. Do not remove clothing if adhering to skin.
- Keep victim calm and warm.

ERG 2016   *Page 171*

Fig. 2–25. ERG—Response guide outlining emergency response information

This section should be used when the entry in the yellow or blue section is *not* highlighted. If product identification cannot be made by any of the index methods, the first responder should refer to Guide 111, which is the first guide in the orange section. The responder should use this guide only until the product is identified.

Green section—initial isolation and protective action distances. The green section should be used to look up the highlighted entries found in the yellow and blue sections. It is designed for materials that present a toxic-by-inhalation hazard. Specific isolation and protective action distances have been established for this group of hazardous materials. This section is organized by ID number.

The highlighted entries in the blue and yellow sections indicate that the material presents a poison and/or poison inhalation risk. If the material is highlighted and *not* on fire, go directly to Table 1 in the *Guidebook*'s green section. If the material is on fire or a fire is involved, also consult the given orange guide page and use the evacuation information under the public safety section. This is a very important instruction to follow because there are some highlighted products that become a greater hazard when involved in fire, and the evacuation distance may need to be increased to the distances noted in the orange guide.

Table 1 in the green section contains a *first ISOLATE in all directions* column (fig. 2–26). This distance indicates the area where an immediate hazard may be present. The next column indicates the *protective action distance*. This information should be used by emergency responders to make recommendations regarding the potentially affected populations. These distances, however, are valid for a period of only 30 minutes following the release. If the spill has passed the 30-minute mark, then the orange guide should be referenced for the evacuation distances. The protective action distances are further broken down based on whether the spill is during the day or at night and also whether the spill is large or small. A small spill is considered 200 L (53 gal) or less, or approximately the size of a 55-gal (208-L) drum. Large spills are considered anything that exceeds this amount.

		SMALL SPILLS				LARGE SPILLS			
		(From a small package or small leak from a large package)				(From a large package or from many small packages)			
		First ISOLATE in all Directions		Then PROTECT persons Downwind during-		First ISOLATE in all Directions		Then PROTECT persons Downwind during-	
ID No.	NAME OF MATERIAL	Meters (Feet)		DAY Kilometers (Miles)	NIGHT Kilometers (Miles)	Meters (Feet)		DAY Kilometers (Miles)	NIGHT Kilometers (Miles)
1815	Propionyl chloride (when spilled in water)	30 m	(100 ft)	0.1 km (0.1 mi)	0.1 km (0.1 mi)	30 m	(100 ft)	0.3 km (0.2 mi)	0.8 km (0.5 mi)
1816	Propyltrichlorosilane (when spilled in water)	30 m	(100 ft)	0.1 km (0.1 mi)	0.2 km (0.2 mi)	60 m	(200 ft)	0.6 km (0.4 mi)	2.0 km (1.3 mi)
1818	Silicon tetrachloride (when spilled in water)	30 m	(100 ft)	0.1 km (0.1 mi)	0.3 km (0.2 mi)	100 m	(300 ft)	0.9 km (0.6 mi)	2.9 km (1.8 mi)
1828	Sulfur chlorides (when spilled on land)	30 m	(100 ft)	0.1 km (0.1 mi)	0.2 km (0.1 mi)	60 m	(200 ft)	0.7 km (0.5 mi)	1.2 km (0.8 mi)
1828	Sulfur chlorides (when spilled in water)	30 m	(100 ft)	0.1 km (0.1 mi)	0.2 km (0.1 mi)	30 m	(100 ft)	0.4 km (0.2 mi)	1.2 km (0.8 mi)
1828	Sulphur chlorides (when spilled on land)	30 m	(100 ft)	0.1 km (0.1 mi)	0.2 km (0.1 mi)	60 m	(200 ft)	0.7 km (0.5 mi)	1.2 km (0.8 mi)
1828	Sulphur chlorides (when spilled in water)	30 m	(100 ft)	0.1 km (0.1 mi)	0.2 km (0.1 mi)	30 m	(100 ft)	0.4 km (0.2 mi)	1.2 km (0.8 mi)
1829 1829 1829 1829 1829 1829	Sulfur trioxide, inhibited Sulfur trioxide, stabilized Sulfur trioxide, uninhibited Sulphur trioxide, inhibited Sulphur trioxide, stabilized Sulphur trioxide, uninhibited	60 m	(200 ft)	0.4 km (0.2 mi)	1.0 km (0.6 mi)	300 m	(1000 ft)	2.9 km (1.8 mi)	5.7 km (3.6 mi)
1831 1831 1831 1831	Sulfuric acid, fuming Sulfuric acid, fuming, with not less than 30% free Sulfur trioxide Sulphuric acid, fuming Sulphuric acid, fuming, with not less than 30% free Sulphur trioxide	60 m	(200 ft)	0.4 km (0.2 mi)	1.0 km (0.6 mi)	300 m	(1000 ft)	2.9 km (1.8 mi)	5.7 km (3.6 mi)

TABLE 1 - INITIAL ISOLATION AND PROTECTIVE ACTION DISTANCES

Fig. 2–26. ERG—Initial isolation and protective action distances

Table 2 lists the chemicals that will produce large amounts of toxic-by-inhalation (TIH) gases when spilled in water (fig. 2–27). This table is also arranged by the materials' ID numbers.

TABLE 2 - WATER-REACTIVE MATERIALS WHICH PRODUCE TOXIC GASES

Materials Which Produce Large Amounts of Toxic-by-Inhalation (TIH) Gas(es) When Spilled in Water

ID No.	Guide No.	Name of Material	TIH Gas(es) Produced
1898	156	Acetyl iodide	HI
1923	135	Calcium dithionite	H_2S SO_2
1923	135	Calcium hydrosulfite	H_2S SO_2
1923	135	Calcium hydrosulphite	H_2S SO_2
1929	135	Potassium dithionite	H_2S SO_2
1929	135	Potassium hydrosulfite	H_2S SO_2
1929	135	Potassium hydrosulphite	H_2S SO_2
1931	171	Zinc dithionite	H_2S SO_2
1931	171	Zinc hydrosulfite	H_2S SO_2
1931	171	Zinc hydrosulphite	H_2S SO_2
2004	135	Magnesium diamide	NH_3
2011	139	Magnesium phosphide	PH_3
2012	139	Potassium phosphide	PH_3
2013	139	Strontium phosphide	PH_3
2308	157	Nitrosylsulfuric acid	NO_2
2308	157	Nitrosylsulfuric acid, liquid	NO_2
2308	157	Nitrosylsulfuric acid, solid	NO_2
2308	157	Nitrosylsulphuric acid	NO_2
2308	157	Nitrosylsulphuric acid, liquid	NO_2
2308	157	Nitrosylsulphuric acid, solid	NO_2
2353	132	Butyryl chloride	HCl
2395	132	Isobutyryl chloride	HCl
2434	156	Dibenzyldichlorosilane	HCl
2435	156	Ethylphenyldichlorosilane	HCl

Chemical Symbols for TIH Gases:

Br_2	Bromine	HF	Hydrogen fluoride	PH_3	Phosphine
Cl_2	Chlorine	HI	Hydrogen iodide	NO_2	Nitrogen dioxide
HBr	Hydrogen bromide	H_2S	Hydrogen sulfide	SO_2	Sulfur dioxide
HCl	Hydrogen chloride	H_2S	Hydrogen sulphide	SO_2	Sulphur dioxide
HCN	Hydrogen cyanide	NH_3	Ammonia		

Fig. 2—27. ERG—Materials producing toxic gases when mixed with water

Miscellaneous entries. Some hazardous materials listed in the ERG, such as styrene monomer, have a P listed after the guide number (128P). The *P* indicates that the material may *polymerize*. A chemical that polymerizes builds on itself, creating a chain of molecules, which could potentially become unstable. Usually an inhibitor is added to a monomer in transportation. Nevertheless, responders should be *extremely cautious* when responding to this type of material.

Shipping papers

Shipping papers are a valuable resource in determining the type of material involved as well as the quantity that responders have to contend with. Shipping papers can provide response personnel with the following information:

- Shipper's name and address

- Receiver's name and address

- List of shipped material(s)
- Proper shipping name of the material
- Emergency response telephone number

Shipping papers are required for all modes of transportation, although the name of the document, the person responsible, and the location differ depending on the mode of travel. When searching for the shipping papers, it may be necessary to check the entire area of each location. In airplanes and trucks, the shipping papers are located near the pilot or driver. In an emergency, shipping paper information can be obtained from the shipper/manufacturer or through CHEMTREC. Table 2–11 shows the various types of shipping papers and their names, locations found, and who is responsible for the papers based on the mode of travel.

After accessing the shipping papers, the first responders will be able to determine if the load contains any type of hazardous materials by looking for the following information:

- First-listed order of hazardous materials on the shipping papers
- "X" in the HM column indicating hazardous materials. Note that "RQ" may replace the "X" in the hazardous materials column if the material is considered a reportable quantity.
- Material's hazard class
- UN/NA ID number

Table 2–11. Types of shipping papers

Mode of Transportation	Title of Shipping Papers	Location of Shipping Papers	Responsible Person(s)
Highway	Bill of lading	Cab of vehicle	Driver
Rail	Waybill/consist	Engine or caboose	Conductor
Water	Dangerous cargo manifest	Bridge or pilot house	Captain or master
Air	Air bill	Cockpit	Pilot

Figure 2–28 is an example of what a hazardous materials entry would look like on a shipping paper.

USER: BE CERTAIN TO IDENTIFY HAZARDOUS MATERIALS BY PLACING AN "X" IN COLUMN 2 BELOW. IN COLUMN 3 INSERT (a.) PROPER SHIPPING NAME, (b.) HAZARDOUS CLASSIFICATION, (c.) IDENTIFICATION NUMBER, (d.) PACKING GROUP IN THAT ORDER. FAILURE TO COMPLY MAY RESULT IN $25,000 FINE AND FIVE YEARS IN JAIL.

DRJ ASSOCIATES • (201) 228-9300

STRAIGHT BILL OF LADING - SHORT FORM - ORIGINAL - Not Negotiable

RECEIVED, subject to the classifications and lawfully filed tariffs in effect on the date of the issue of this Bill of Lading.

SHOW THIS NUMBER ON FREIGHT BILL

CARRIER: **NOVA Shipping**

AT: DATE: 7/25/01   FROM: ABC Chemical Company, Inc.   SHIPPER'S NO. 546595656

PERMANENT POST-OFFICE ADDRESS OF SHIPPER

ABC CHEMICAL COMPANY, INC., ANYWHERE, USA 26598

CARRIER: IF THIS IS A PREPAID SHIPMENT ATTACH FREIGHT BILL AND MAIL BOTH DOCUMENTS TO:

CONSIGNED TO **Santoro's Cleaners**   STREET ADDRESS OF CONSIGNEE **1256 East Main Street**   CAR OR VEHICLE INITIALS / NO.

DESTINATION CITY **Torrington**   STATE **CT**   ZIP **06790**   REQUIRED ON-TIME DELIVERY DATE IS

FOR HELP IN CHEMICAL EMERGENCIES INVOLVING SPILL, LEAK, FIRE EXPOSURE CALL CHEMTREC 800-424-9300 TOLL FREE ANYTIME

NUMBER OF PIECES & TYPE	H M	DESCRIPTION OF MATERIALS, SPECIAL MARKS AND EXCEPTIONS	WEIGHT SUB TO CORR	LB/GAL
5	X	15 GAL SSDD TETRACHLOROETHYLENE, 6.1, UN-1897, PG III,	1,000 lbs	
		RQ, MARINE POLLUTANT, NA EMERGENCY GUDE NO. 160		

Fig. 2–28. Shipping paper example

INITIATE PROTECTIVE ACTIONS

Isolate area and deny entry

HM 4.3.1 Based on the incident conditions protective actions can range from blocking roads to having victims remove themselves from the immediate hazard area. Once a hazardous material has been recognized, the first responder should first determine the initial isolation distance based on the ERG. They should make every effort to ensure all persons in the immediate area fall back to at least the determined distance. Since awareness level personnel are not trained or equipped to perform formal evacuations this must be accomplished through verbal directions to all those in the immediate area. Additional information can be ascertained from shipping papers, SDS or online apps such as WISER. Once all endangered people have fallen back, awareness level personnel should make all attempts to isolate the area and deny entry. This is best accomplished by blocking roads and access points surrounding the incident (fig. 2–29). Everyone not directly involved in emergency response operations must be kept away from the area. Unprotected emergency responders are not allowed entry into the isolation area. This "isolation" task is done to establish control over the area of operations. It is the first step for any protective actions that may follow.

Fig. 2–29. One of the first steps at a hazmat incident is to isolate the area and deny entry. Awareness level responders can perform this task.

NOTIFICATIONS

Requesting additional resources

HM 4.4.1 Notifications for the first responder trained to the awareness level may simply be notifying the fire department, while notifications at hazardous materials incidents can become complex. However, based on agency policies and SOGs they may be required to notify available resources such as hazmat response teams, mutual aid fire, police, and EMS agencies and environmental protection agencies, and so on (fig. 2–30). Government agencies such as landfill or public works personnel should also understand each department or agency's response objectives and the role they play in the hazmat responses in their jurisdiction. A list of these available resources should be established, maintained, and included with the resource listing in the LERP.

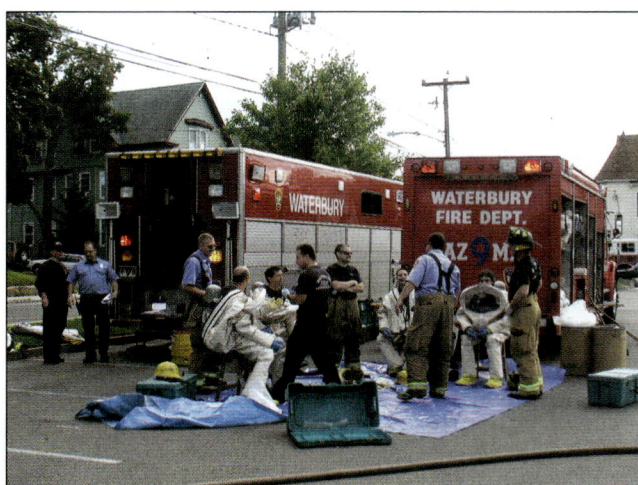

Fig. 2–30. First responders should be familiar with available resources such as a hazmat team.

QUESTIONS

1. List three types of fixed facilities that may house hazardous materials.

2. NFPA 704 placards are usually found where?

3. The Hazardous Materials Information System (HMIS) marking used by industry is similar to NFPA 704 except for an additional section that was added. That section is to identify what?

4. The 4-digit number placed inside of a DOT placard is known as what?

5. How many hazard classes are there in the DOT placarding system used for tank trucks, rail cars and other vehicles transporting hazardous materials?

6. What are the main purposes or goals of the Global Harmonization System (GHS)?

7. What do the green section/pages of the DOT *Emergency Response Guidebook* tells first responders?

8. Shipping papers used for rail operations are known as _____ and are kept in either the_____ or the _____.

9. The best and most immediate two actions for an Awareness Level responder to take upon arrival at a scene are to: _____ and _____

10. The blue section/pages of the DOT *Emergency Response Guidebook* tells first responders what?

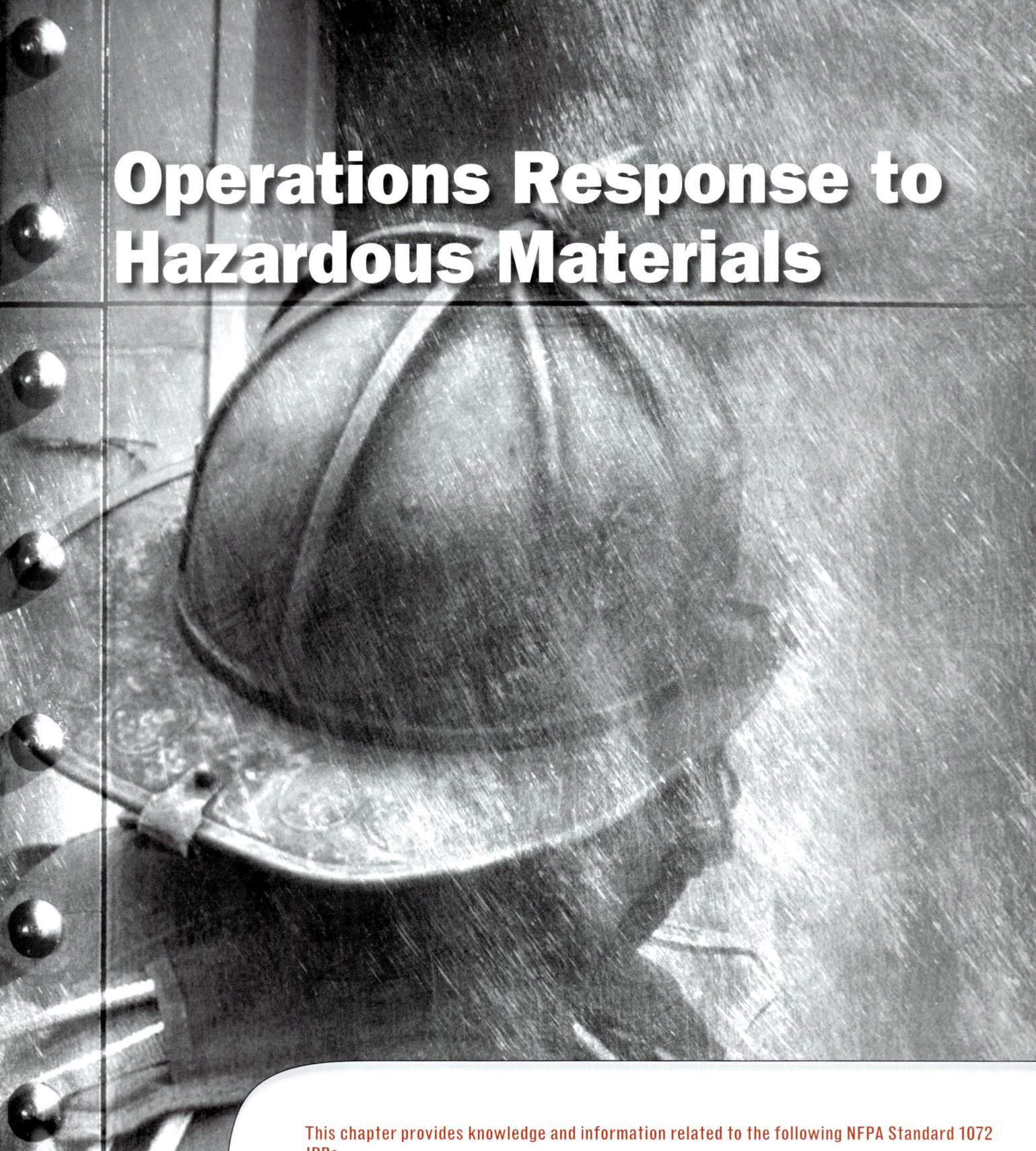

Operations Response to Hazardous Materials

This chapter provides knowledge and information related to the following NFPA Standard 1072 JPRs.

5.1.1	5.4.1
5.1.5	5.5.1
5.2.1	5.6.1
5.3.1	

OBJECTIVES

Upon completion of the chapter, you should be able to do the following:

- Scene size-up and determine conditions and scope on arrival
- Understand chemical and physical properties
- Understand the use of written and electronic reference materials
- Identify vehicles that transport hazardous materials, e.g., tank truck, trailers, And rail cars
- Identify storage containers, tanks and vessels.
- Understand basic fire protection in fixed facilities.
- Make contact with shippers, manufacturers and agencies to share and obtain information
- Understand and recognize the signs of terrorism and criminal activity
- Understand how the Incident Command System works for hazardous materials incidents
- Understand the routes of exposure in to the human body
- Action planning for evacuation and shelter in place
- Operational procedures; offensive vs. defensive vs. withdrawal
- Explain the procedure for gross and emergency decontamination

IDENTIFYING POTENTIAL HAZARDS

Surveying the scene

HM 5.1.1 Surveying the scene is arguably the most important phase because it forms the basis for all decisions made for the remainder of the incident. Effective decisions can be made only if the effort is made to collect all available information on a release and then consult available resources to determine the correct actions (fig. 3–1). Great care must be taken during this process because something as simple as misspelling the chemical name can result in major problems.

Fig. 3–1. Effective decisions require collecting information on the product released.

The three basic steps in this phase are:

- Surveying the scene to collect clues on the type(s) of materials involved

- Collecting the hazard and response information regarding those materials

- Determining the behavior of the material based on what the physical properties are determined to be

Surveying the scene can be done in conjunction with the *isolate the area and deny entry process* that follows.

Using the senses

HM 5.1.5 Upon arrival on the incident scene, the responder's senses provide the first clues that the incident involves some type of hazardous materials release. Of the five senses, sight and sound are the only ones that can be used safely. Sight and sound can provide valuable clues to response personnel. For example, liquids pooling on the floor or the ground, chemical reactions taking place, a vapor cloud or fire, injured persons in the area, condensation lines on pressurized tanks, corrosive reactions taking place, and the boiling of unheated substances are all visual indicators that a release has occurred. The hissing sound of a pressurized container releasing its contents or the banging/creaking from overstressed containers are other indicators of a pending release or a

release that has occurred. While smells can be used to identify materials such as the obvious smell of a natural gas leak, using the sense of smell should be discouraged. The sense of taste and touch is strictly prohibited in the identification process.

Unfortunately, using the senses is limited in that the closer a first responder gets to the hazmat incident to identify the product, the greater the risk of exposure. Also, many materials are odorless, tasteless, and colorless, thus providing no warning to those being exposed or it is possible for some materials to deaden the senses so that, although it is still present in the air, it cannot be smelled by anyone in the area. It is always best to conduct scene surveys from a distance using binoculars or other visual aids to stay out of the exposure area (fig. 3–2).

Fig. 3–2. Always maintain a safe distance when collecting hazard data and information.

Scene conditions

HM 5.2.1 Good observation skills play a key role on the initial response and size-up. If responders fail to take a good look at the incident as a whole, they often miss important pieces of information. As with any emergency scene, tunnel vision can be dangerous and must be avoided at all costs. When arriving on scene, consider some of the following questions and make note of what is occurring or will potentially occur:

- *What type of container is involved?* Is it a rail car, cargo tank, intermodal, or fixed facility tank?

- *What is the container size?* Is it a small cardboard box, a 55-gal (208-L) drum, or a 50,000-gal fixed-facility bulk tank and how much potential release of product is there?

- *Is the container marked?* Are there visible markings such as DOT placards, UN/NA numbers, NPFA 704 symbols or product names and ID numbers that can be used to determine the product or general hazards (fig. 3–3)?

Fig. 3–3. Container markings can help identify the product and/or its hazards.

- *What is the physical state of the material?* Is it a solid, liquid, or a gas?

- *How badly is the container damaged?* Is it a large tear, a small hole, or broken piping or valving, and what is the rate of product release?

- *How long has the container been leaking?* Is it a few minutes, several hours, or a few days, and how much product was potentially released?

- *Is the material on fire?* Is extinguishment necessary, or would it be better to let the material burn off?

- *Are there any injuries involved?* Are they minor or serious? Will victims need emergency decontamination (fig. 3–4)?

- *What is the topography like?* Is it flat, allowing for minimal product spread, or are there hills, facilitating the flow of the product away from the source of the leak?

- *How is the land used?* Is it industrial, farming, residential, or a highway, and how will the release potentially impact the area?

- *How accessible is the release?* Will response personnel and vehicles have access to the scene, or will extra equipment be needed?

- *What are the weather conditions?* Is it hot or cold? Is it raining or snowing? What is the wind speed and direction? How do these conditions impact the spread of the material?

- *Are there any bodies of water nearby?* Is there a danger of the material reaching a reservoir, lake, pond, river, or stream, thus spreading the contamination?

- *What, if any, are the public exposure risks?* Was the release in an urban or rural area, and how much of the population is at risk (fig. 3–5)?

- *What types of utilities are in the area?* Are there any above- or below-grade wires or pipelines in the area that could provide a potential ignition source or a path for the material to flow along?

Fig. 3–4. If there are any victims, determine if they require emergency decontamination.

Fig. 3–5. Hazmat releases in urban areas can provide greater challenges than in rural areas because of the dense population.

Operations Response to Hazardous Materials

HAZMAT

Chapter 3

- *Are there any storm or sewer drains nearby?* How close is the product to entering the drains and where do those drains lead?

- *Do potential sources of ignition exist?* Are any of the following present in the area such as lighting equipment, chemical reactions, electric motors and controllers, open flames, cutting or welding operations, smoking materials, or portable heating equipment?

- *Is the incident inside a structure?* Are there people inside of the building? Are there any floor drains or an HVAC system involved that could inadvertently lead to a spread of the material?

- *Is anything out of the ordinary occurring?* Is there evidence that this release could involve some type of criminal or terrorist activity based on the material and location or by visible clues on scene?

After making an initial observation of the scene, the firefighter can obtain a better idea of the incident scope and determine whether additional resources are needed, or if the first-due assignment can properly mitigate the situation. Additionally, the responder can determine the number of potential exposures in the area and establish other valuable pieces of information that can be used when planning the response phase.

Determining the behavior of the material

A couple factors play into figuring out what the material is going to do. These factors are identifying the material, assessment of the container and its contents and the physical chemical properties of the material. When the responder has identified the material, the next step is to begin to estimate its likely behavior as it is released from the container. The type of damage to the container, topography, and environmental conditions are just a few of the additional factors that affect how the material will behave during the release. With this information it is also possible to begin to estimate the potential harm resulting from the leak.

CONTAINER BREACHES

In the mid 1970s, Ludwig Benner Jr. created the General Emergency Behavior Model (GEMBO) with the intent of helping responders limit the effects of a hazardous materials release. This model is designed to aid in the threat prediction process and accounts for the various factors involved in predicting the behavior of a release. It is broken down into the following sections: stress, breach, release, engulf, contact, and harm.

Stress

Different types of stresses can be applied on the container, depending on the situation. It is also possible to have more than one of the following stresses acting on a container at any given time. For example, heating a container may weaken it and cause a chemical reaction to take place, or a mechanical stress may cause damage to a container and thermal stress can initiate a chemical reaction and/or cause the container pressure to rise. The following are examples of common types of container stressors.

Thermal stress. Thermal stress can be the result of radiated, convected, conducted, or direct heat exposure (fig. 3–6), as well as exposure to cold. Heat or cold can result in the expansion, contraction, and weakening of the container. Thermal stress may increase internal pressure or initiate a chemical reaction while reducing container strength simultaneously, resulting in sudden failure.

Fig. 3–6. Fire can place a thermal stress on a container.

Always be on the lookout for stressed containers when responding to a hazardous materials incident. If the material has been released before arrival, not only should responders address that release, but also determine the potential for other containers in the area to be stressed. Always be on the lookout for indications of thermal stress, such as the operation of a safety relief valve or the bulging of a container.

When heat is applied to a closed container, a boiling liquid expanding vapor explosion (BLEVE) may occur. This starts with the liquid inside boiling and exerting pressure on the container. The internal pressure increases until the container fails, which may result in an explosive release of the material (fig. 3–7).

Fig. 3–7. A BLEVE can result in catastrophic container failure.

The release of pressure may result in a violent rupture, causing fragments of the container to be projected significant distances. There is also a shock wave associated with a BLEVE. If the liquid is flammable, the container failure will be accompanied by a fireball.

Mechanical stress. Mechanical stress is the physical application of energy that results in container or attachment damage. It may cause the container to become deformed, cracked, penetrated, lose some of its thickness, or damage valves or piping (fig. 3–8). Mechanical stress is the result of a collision between two objects such as two vehicles colliding or a vehicle colliding with the road.

Fig. 3–8. Containers can become damaged due to mechanical stress. Note the stress crack just left of the labeling.

The following are indications of mechanical stress:

- The container may become deformed or penetrated.

- It may also be punctured, gouged, scored, broken, or torn.

Responders should be aware that it is possible for a container to be damaged or weakened by a mechanical force without any visible signs of a potential release.

Chemical stress. Chemical stress results from an uncontrolled reaction or interaction of the container and/or its contents. The onset of container deterioration may occur suddenly or over a long period of time. Chemical reactions taking place in a container may also result in excessive heat and/or pressure that may result in container failure. The following are indicators of chemical stress:

- Containers that display visible corrosion

- Pressure or heat resulting from the reaction of the substances

Breach

After the container has been stressed to the point where a breach occurs, responders can expect the container to fail in one of the following manners.

Disintegration. Disintegration results in the total loss of container integrity. An example of this type of breach is a glass container shattering (fig. 3–9).

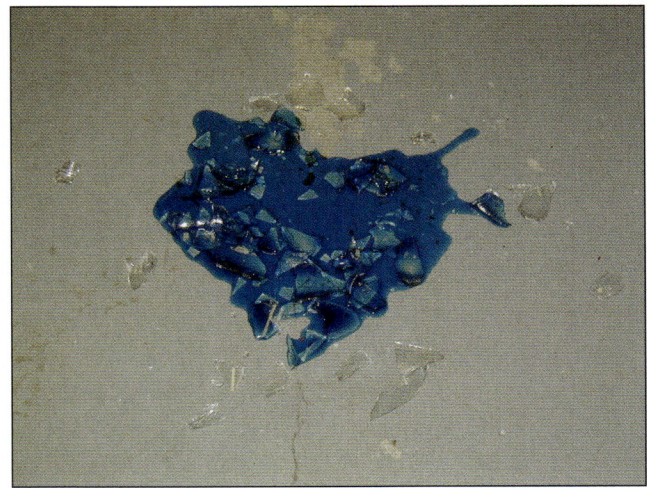

Fig. 3–9. Disintegration is the complete failure of a container.

Runaway cracking. Runaway cracking is the rapid expansion of a small crack in a closed container. The result is a container violently breaking into two or more pieces. This type of breach generally occurs in closed containers such as liquid drums or pressure vessels.

Attachment failure. In this type of breach, attachments may open up or break off. For example, safety relief valves, fusible plugs, discharge valves, and so on, may break.

Puncture. Punctures are generally caused by mechanical stresses and result in container failure. For example, a forklift may pierce a 55-gal (208-L) drum (fig. 3–10).

Fig. 3–10. A puncture in a container can result in the release of a material.

Split or tear. A split or tear is commonly associated with fiber drums or plastic bags and can also refer to, for example, split 55-gal (208-L) drums and seam or weld failures.

Release

After the container becomes breached due the stresses placed on it, the material can be released in several different ways depending on the material involved and other factors. The following are examples of different types of releases.

Detonation. Detonation is an instantaneous and explosive chemical reaction that releases pressure and extreme heat. It results in the fragmentation, shattering, or disintegration of the container. Detonation occurs in 1/100 of a second or less; the speed of this reaction affords first responders no time to react. Examples of materials that can detonate include military ammunition, mining explosives, and organic peroxides.

Violent rupture. Violent ruptures are chemical reactions that have a release lasting less than one second. It is also known as deflagration. It is associated with runaway cracking and overpressurization of closed containers. Containers and their contents may become projectiles as a result of the reaction. Like detonation, the speed of violent rupture affords first responders no time to react.

Rapid relief. Rapid relief is the fast release of a pressurized hazardous material. It may take several seconds to several minutes to release. The speed of the rapid relief depends on the size of the opening in the container, the properties of its contents, and the type of container.

In the case of rapid release, the pressure is released through:

- Relief valves
- Broken or damaged piping/valves
- Punctures
- Tears or splits

These types of releases typically give time for first responders to react and move to a safe location.

Spills and leaks. Spills and leaks result in the slow release of hazardous materials. The duration of release varies from minutes to days. Slower release rates from spills and leaks allow adequate time to develop countermeasures. Spills and leaks are nonviolent low-pressure flows resulting from the following:

- Broken valves/piping
- Tears, splits, or punctures in a container

Engulfment

After the material is released from its container, it begins to engulf the area around it. The extent to which this occurs depends on the amount of material, the physical state of material (solid, liquid, gas), weather conditions, topography, type of container breach, control efforts of emergency responders, and the type of release and its duration. The following are examples of different engulfing events.

Hemisphere. The hazardous material is airborne in a semicircular or dome shaped pattern (fig. 3–11). It may be partially in contact with ground or water. The hazardous material initially rises and spreads outward symmetrically and falls equally in all directions. This type of event generally occurs when the winds are calm.

Cloud. The hazardous material is airborne in a ball-shaped pattern (fig. 3–12). The material rises above the ground or water.

Plume. The hazardous material is airborne in an irregular pattern (fig. 3–13). The material is affected by the materials vapor density and drifts with the wind. The trajectory of the substance is determined by wind and topography.

Fig. 3–11. A hemisphere engulfing pattern

Fig. 3–12. A cloud engulfing pattern

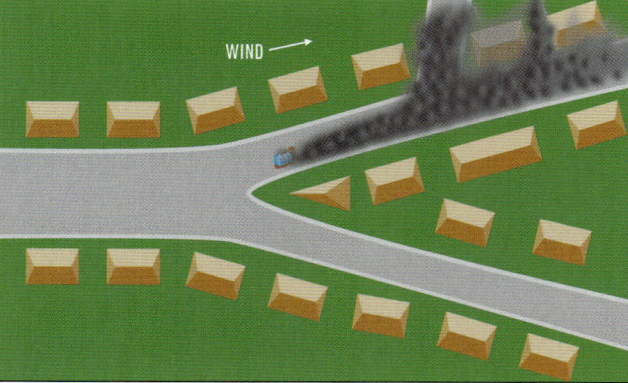

Fig. 3–13. A plume engulfing pattern

Cone. The hazardous material is airborne in a triangular pattern flowing away from the release point (fig. 3–14). The material flows with the topography and widens as it gets further away from the source.

Fig. 3–14. A cone engulfing pattern

Stream. The hazardous material follows a path determined by gravity and ground topography (fig. 3–15). It can be a liquid or a gas.

Fig. 3–15. A stream engulfing pattern

Pool. A liquid hazardous material follows a flat, circular pattern (fig. 3–16). The spill spreads equally in all directions and may be present on the ground or on the surface of a body of water.

Fig. 3–16. A pool engulfing pattern

Irregular. Hazardous materials are deposited in an irregular pattern (fig. 3–17). For example, materials may be carried by contaminated personnel.

Fig. 3–17. An irregular engulfing pattern

Contact

As the released material begins to engulf the area around the container, a contact event can occur. A contact event refers to the amount of time an exposure (person or object) is in contact with the hazardous material. The factors influencing the duration of contact are the quantity of material released, the dispersion method, and the speed of release. Responders can use the ERG to create a basic plume model based on isolation distance and downwind protection distances. There are more sophisticated plume modeling apps and software that are available to responders. There are four general time periods associated with a contact event: immediate, short-term, medium-term, and long-term.

Immediate event. In an immediate event, durations are measured in seconds and milliseconds. Examples include explosion, detonation, or deflagration.

Short-term event. A short-term event has a duration of minutes to hours. The event is characterized by a lower hazard, small (short) releases, and is quickly brought under control. Examples include transient vapor clouds.

Medium-term event. A medium-term event occurs over a period of days, weeks, or months. The event is characterized by a moderate to high hazard. First responders have a lengthy or difficult decontamination and moderate cleanup is necessary. Examples include lingering pesticides resulting from a fire or spill.

Long-term event. A long-term event extends over years or generations. It results in heavy contamination

and has a lengthy and difficult decontamination. In the case of a long-term event, extensive cleanup is required. Examples include permanent radioactive sources such as Chernobyl.

Harm

Harm is the injury or damage resulting from exposure to the hazardous materials release and subsequent exposure. The three factors that determine the potential harm caused by an exposure are as follows:

- Timing of release: refers to the speed at which the material is released as well as the amount of time spent in contact with the material based on the movement of the material

- Size of release: refers to the size of the area which is covered by the material

- Hazards of the material: refers to the relative level of harm that exposure to it causes

SCENE CONTROL

HM 5.4.1 **Scene control** operations are designed to protect both the public and responding personnel. To protect the public, responders should isolate the area, deny entry, and begin to institute protective actions such as protect-in-place or evacuation. In addition, to provide protection for emergency responders, access should be limited to the area, and control zones (hot, warm, and cold) should be identified and established.

Scene control priorities

Upon arrival on scene and after the initial size up determines that a hazmat incident is involved, responders should immediately begin the process of determining the size of the endangered area, isolating that area, and denying entry. This means that anyone not directly involved with the response should be removed from the immediate area, and civilians and unprotected responders should not be allowed into the isolation area. Determining the size of the endangered area is most easily accomplished by using the Department of Transportation Emergency Response Guide (DOT ERG) (fig. 3–18). As more information is gathered about the product, decisions will need to be made regarding the additional protective actions that should be used (protect-in-place and/or evacuation) to protect the public, and control zones (hot, warm, and cold) will need to be set up for the protection of emergency response personnel.

Fig. 3–18. The DOT ERG can be used to determine the size of the endangered area.

Public protective action options

HM 4.3.1 Protecting the public will depend largely on the fire department and other responders and the local emergency planning committee (LEPC) to develop an effective emergency plan. This plan needs to have a risk analysis of the potentially affected population. Because responders in the early stages are small in number and possibly overwhelmed, the need for accurate and up-to-date information is essential in developing an action plan that will include steps needed to protect the public.

When a hazardous material event occurs, the following options need to be considered when developing a plan to protect the public. Operations level responder can and should perform these tasks:

- Isolate and deny entry
- Evacuation of threatened populations
- Protect-in-place for all at-risk populations
- A combination of evacuation and protect-in-place

Additional information pertaining to evacuation and protect in place is covered later in this chapter.

IDENTIFYING POTENTIAL HAZARDS

Physical and chemical properties—terms and definitions

For responders to make educated decisions about the materials involved in a hazmat incident, they must understand the properties that govern the material's behavior. For example, a chemical that could present a significant hazard when released on a hot summer day may be relatively inert on a subzero winter day. Understanding the physical properties of the material will allow operations-level responders to develop an appropriate and safe defensive response plan based on available personnel and equipment. During the information collection process, it is important that personnel understand the data provided by the various research resources that will be discussed later on in this chapter. The hazmat responder will likely come in contact with the following terms.

Chemical reactivity

Chemical reactivity is the ability of a material to undergo a chemical reaction with another substance. During the reaction process, increased temperature and pressure buildup can result in catastrophic failure of a container.

Exothermic and **endothermic** are two types of chemical reactions. The first type is a reaction that causes a release of heat. Exothermic reactions occur with most chemical reactions. An endothermic reaction is a chemical reaction that is accompanied by the absorption of heat. An example of this is the reaction that takes place inside an ice pack.

Unstable materials can react violently with little or no outside interaction. Unstable substances are those capable of undergoing spontaneous change. For example, nitroglycerin may explode if jarred, heated, or otherwise contaminated. Examples of unstable materials include compounds that crystallize or deteriorate such as picric acid, ether, organic peroxides, and dynamite.

Hypergolic materials ignite when they come in contact with each other. Hypergolic reactions occur when a fuel and an oxidizer are mixed.

Pyrophoric materials are those elements that react and ignite upon coming in contact with air.

Corrosivity (pH)

Corrosives are substances that destroy or burn living tissues and have destructive effects on other types of materials. Containers can be damaged and lose structural integrity if they are not designed to contain corrosive materials.

Corrosives are divided into acids and bases (alkaline).

Contact with acids can cause severe chemical burns and permanent eye damage. Usually a person who comes in contact with an acid feels pain. Common acids include hydrochloric, nitric, and sulfuric. All acids have a pH value of less than 7. Another property of acids is that they dissolve metals and react with bases.

Bases break down fatty skin tissues and penetrate deeply into the body. Coming in contact with a base does not normally result in immediate pain. Rather, a greasy or slick feeling on the skin is indicative of an exposure. This is the result of the base breaking down the fatty tissue in the skin. Common bases include caustic soda, potassium hydroxide, and other alkaline materials. A base has a pH value greater than 7.

Particle size

Particle size is a term that refers to the size of a solid particle and is usually measured in microns.

Persistence

Persistence refers to the ability of a material to remain in the area of release for long periods of time. It is intended to prevent responders from reentering an area in which concentrations of the material remain high for an extended period of time.

Physical state

Physical state refers to the form a material exists in: gas, liquid, or solid. **Gases** are substances that have no definitive shape or volume. **Liquids** are substances that have no definite shape, but do have a specific volume. **Solids** are substances that have both a specific shape and volume. The severity of the release and methods of control vary based upon the physical state of the product. Solids tend to be the easiest to control because they remain in place unless acted upon. Liquids are more difficult to control than solids because they flow with gravity. Gases are the hardest and may require extensive evacuations.

The physical state of a material can also determine the most likely method of exposure. Solids generally enter through contact, ingestion, or the inhalation of particles such as dust. Liquids are typically ingested, absorbed through the skin, or inhaled if they are evaporating. Gases can be inhaled, absorbed through the skin, and possibly ingested.

Specific gravity

Specific gravity is the weight of a substance compared to the weight of an equal volume of water at a given temperature (fig. 3–19 a and b). It determines whether a substance sinks or floats in water. Water is given a value of 1; materials with a value greater than 1 will sink in water, and materials with a value less than 1 will float.

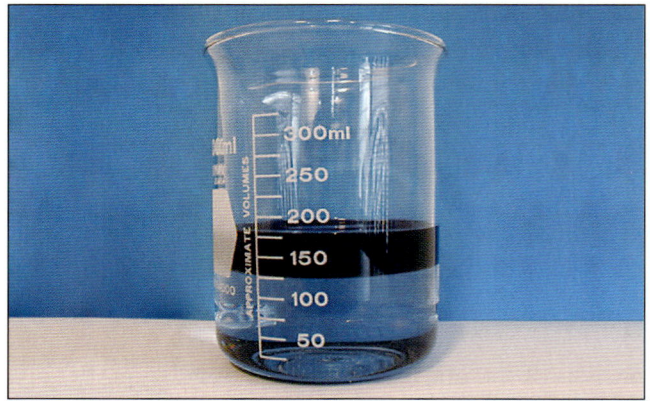

Fig. 3–19a. Materials with a specific gravity less than 1 will float in water.

Fig. 3–19b. Materials with a specific gravity more than 1 will sink in water.

Specific gravity is important when determining the effectiveness of pads and booms. Most hydrocarbons have a specific gravity of less than 1 and will float on water. Materials that sink are more difficult to contain and clean up. Water solubility is a concern with materials that have a specific gravity greater than 1.

Water solubility

Water solubility (also known as miscibility) is the ability of a liquid to mix with water. It is difficult to remove water-soluble materials from a body of water. Corrosives, poisons, and alcohols tend to be water soluble. Water-soluble materials are difficult to extinguish when burning.

Vapor density

Vapor density is the density of a gas compared to the density of air (fig. 3–20a and b). It determines whether a vapor will rise or fall. Air is given a value of 1; gases with a vapor density greater than 1 will sink, and those with a vapor density of less than 1 will rise. For example, propane is a gas with a vapor density greater than air. With a vapor density of 1.56, it will accumulate in low spots; potential ignition sources should be considered.

Fig. 3–20a. Gases with a vapor density less than 1 will rise.

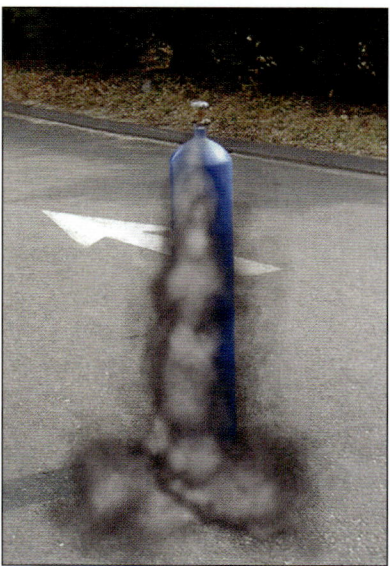

Fig. 3–20b. Gases with a vapor density greater than 1 will sink.

Vapor pressure

Vapor pressure is the pressure at which a liquid and its vapor are in equilibrium at a given temperature. Liquids with high vapor pressures evaporate rapidly. It is important that the first responder consider a material's vapor pressure when deciding evacuation routes or other tactical plans. The higher the vapor pressure, the more dangerous the chemical. Higher vapor pressures also translate into lower boiling points. Chemicals with high vapor pressures emit vapors at lower temperatures than chemicals with low vapor pressures. Examples of substances with high vapor pressures include gasoline, acetone, and alcohol. Substances with low vapor pressures include motor oil, diesel fuel, and water.

Boiling point

Boiling point is the temperature at which a liquid changes to a gas at normal atmospheric pressure; reaching the boiling point results in a rapid change from a liquid to a gaseous state. The lower the boiling point, the more volatile a material is. A liquid with a low boiling point has a high vapor pressure and a low flash point. At temperatures below the boiling point, evaporation takes place only from the surface of the liquid; during boiling, vapor forms within the body of the liquid.

Flash point

Flash point is the minimum temperature at which a liquid gives off sufficient vapors to create an ignitable mixture with air near its surface. At its flash point, a fuel source gives off sufficient vapors to ignite momentarily and will not continue to burn unless additional heat is introduced. The flash point is important in determining the temperature at which a flammable liquid is giving off sufficient vapors to ignite. A liquid with a low boiling point has a high vapor pressure and low flash point.

Ignition temperature

Ignition temperature is the minimum temperature that a fuel in air must be heated to in order to ignite without an independent ignition source being present. It is also known as the autoignition temperature. The ignition temperature of a substance is significantly higher than its flash point. All flammable materials have an ignition temperature, and it is an important factor when determining the potential of ignition of a flammable substance.

Flammable (explosive) range

Flammable (explosive) range is the percentage of the gas or vapor concentration in air that will burn if an ignition source is introduced. The flammable (explosive) range has the following limiting concentrations:

- Lower explosive limit (LEL): if the concentration is below the LEL, it is too lean to burn.

- Upper explosive limit (UEL): if the concentration is above the UEL, it is too rich to burn.

These measurements are given as a percentage in air (fig. 3–21). Concentrations between the upper and lower limits will burn rapidly if ignited, and any concentrations above 10% of the LEL must be considered to have potential for ignition. Containers can be damaged by an explosion if the contents are in their flammable range and an ignition source is introduced. Responders should always consider that ventilating an area above the UEL will eventually bring it back down into the material's flammable range. Caution should always be taken to control potential ignition sources when dealing with flammable atmospheres.

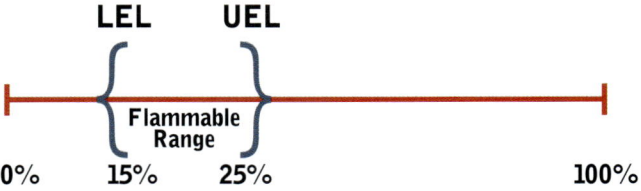

Fig. 3–21. The flammable range of anhydrous ammonia

Toxic by-products of combustion

Toxic by-products of combustion are given off by all substances when they are burning. Responders should keep in mind, however, that substances classified as hazardous materials will often produce more toxic by-products than a fire burning natural materials such as wood. There is also the potential that the hazardous materials could be consumed in the combustion process. The first responder should research the material to determine how it will react under fire conditions.

Radiation

Radiation is energy released in the form of particles or electromagnetic waves. There are two types or groupings of radiation: ionizing and nonionizing.

Ionizing radiation has enough energy to cause a physical change to atoms by triggering them to lose electrons and become **ions**. When first responders refer to radiation in a hazardous materials response, they are usually referring to ionizing radiation. Examples include alpha, beta, gamma, and neutron (fig. 3–22).

Alpha particles are relatively heavy and slow moving. They have poor penetrating power and can be stopped by a piece of paper or skin. Alpha particles consist of two protons and two neutrons.

Beta particles are fast and can move at speeds of 9/10 the speed of light. They can penetrate up to 1 to 2 cm of water or human flesh. Beta particles can be stopped by a sheet of aluminum a few millimeters thick. Because they consist of an electron, they have very little mass compared to an alpha particle.

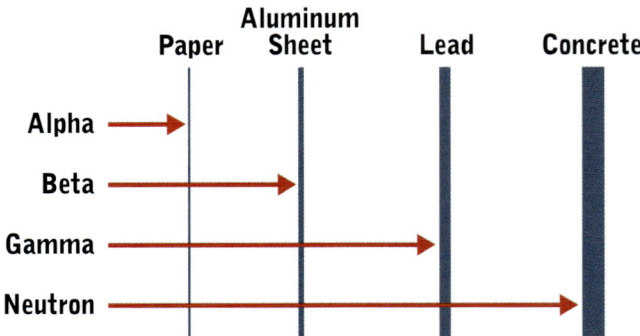

Fig. 3–22. Alpha, beta, gamma, and neutron radiation require different levels of shielding.

Gamma rays are electromagnetic rays that move at the speed of light and can kill living cells. Thick barriers of concrete, lead, or water are used as shielding materials for gamma rays.

Neutron radiation consists of penetrating particles and is much harder to shield against than alpha, beta, and gamma radiation. This type of radiation is not as common as the other types discussed, and it primarily results from the fissioning or splitting of certain atoms within a nuclear reactor. Large quantities of water and concrete are most commonly used to shield the reactor against neutrons. Unlike other types of radiation, neutron radiation can cause some materials, such as alloy steels, to develop an induced radioactivity after absorbing neutron energy. However, the absorption of neutron radiation does not always lead to a radioactive atom.

Nonionizing radiation has enough energy to cause molecules to move or vibrate, but not enough to change them chemically. Examples include radio waves, microwaves, ultraviolet and visible light, and magnetic fields.

HEALTH HAZARDS

Because the top priority of all hazmat responders is the preservation of life, it is important to understand the effects a product may have on human physiology. As such, responders must be familiar with the terminology used to describe health hazards, the effects different classifications of materials have on the body, typical routes of entry, and the threshold limits of potential exposures. This information plays a crucial role in making informed decisions about the incident response and understanding the level of danger faced by those in the exposed area.

Health hazard terminology

When collecting response information it is also important that responders understand the different terms used to describe health hazard concerns. For example, it is important to understand the difference between acute and chronic effects of a material. Responders should be familiar with the following terms and be able to describe the differences between them.

Infections vs. Contagious. Contagious substances can be transmitted between a person or animal to another through some manner of contact, typically through bodily fluids. Infectious refers to a type of disease that is the result of exposure to a harmful microorganism. This microorganism reproduces within the body and begins to attack its organs and cells. Just because a disease is infectious does not necessarily mean it is contagious.

Acute effects vs. chronic effects. Acute effects will present their symptoms immediately, whereas chronic effects can take a substantially longer time to appear, even years after the exposure.

Acute exposures vs. chronic exposures. An acute exposure is typically considered a large dose in a short period of time, whereas a chronic exposure occurs when someone is exposed to a lower dose over a longer period such as days, weeks, or years. Acute exposures can often have both short-term and long-term effects, depending on the substance involved.

Health hazards associated with a hazardous materials release

Responders should have a general idea of the health hazards associated with a release, even before they may know the exact name of the material. Using the basic hazard class information, as well as the physical attributes of the material, certain dangers and health effects can be considered. Additionally, physical hazards associated with the container and working in appropriate personal protective equipment must be assessed. The following are health hazards that may be encountered at a release.

Thermal—heat. Responders should be aware that operating in protective clothing can cause a body to overheat. Exposure to heat can lead to one or more of the following:

- Heat cramps result from heavy exertion and exposure to high heat. They are caused by dehydration and excessive loss of salt in the extremities.

- Heat exhaustion is a mild form of traumatic shock resulting from prolonged physical exertion in a hot environment when the body becomes incapable of releasing excess heat.

- Heatstroke is a serious medical condition that can become life-threatening if it is not treated promptly.

Table 3–1 lists the symptoms of each heat-related condition.

Table 3–1. Symptoms of heat-related conditions

Heat Cramps	Heat Exhaustion	Heatstroke
Heavy perspiration	Heavy perspiration	No perspiration
Skin moist to the touch	Skin cool, pale, and moist	Hot, dry skin
Physical weakness	Dizziness	Weakness and confusion
Muscle cramps	Mildly elevated temp	High body temp
	Weak pulse	Rapid pulse
		Convulsions
		Shallow breathing
		Headache

To prevent overheating, personnel should take preventative measures such as drinking plenty of fluids, utilizing mobile showers or misting setups, setting up rest areas in the shade or an air-conditioned area, rotating personnel, and avoiding coffee and caffeinated drinks (fig. 3–23). If response personnel display any symptoms of these heat-related injuries, they should be removed from the response operation and treated according to local medical protocols.

Fig. 3–23. Response personnel should drink plenty of water to prevent heat-related injuries.

Thermal—cold. Responders have the potential for cold-related injuries when dealing with cryogenic or liquefied gases such as liquid oxygen (LOX), hydrogen, nitrogen, and liquid natural gas (LNG).

These materials can freeze objects and human tissue instantaneously upon contact. Clothing saturated with a cryogenic material must be removed immediately because many of these liquids are flammable, and their vapors pose a serious flammability problem.

Table 3–2 lists the conditions that can result from exposure to cold.

If response personnel display any symptoms of these cold-related injuries, they should be removed from the response operation and treated according to local medical protocols.

Mechanical hazards. Mechanical injury may occur if the responder is struck by an object or rubs up against one. The failure of a pressurized container may also result in serious mechanical injury. The failure of such a container can result in serious bruises, lacerations, or puncture wounds.

Friction injuries are most commonly caused by contact between the skin surface and protective clothing,

causing blisters and brush burns. Friction injuries are less common than striking injuries in hazardous materials incidents.

Poisons. Poisons can cause death or serious bodily injury to responders. There are many different types of poisons, each having certain effects on the human anatomy (table 3–3). The type of damage caused to the body depends on the poison involved in the incident.

Table 3–2. Symptoms of cold-related conditions

Superficial Frostbite	Frost Nip
Waxy or white skin	Whiting or blanching of skin
Outer layers of skin are firm to the touch	
Tissue underneath is still flexible	
Deep Frostbite	**Hypothermia**
Cold and pale skin	Shivering
Skin hard to the touch	Slow pulse and respiratory rate
	Sleepiness, apathy, or listlessness
	Glassy eyes
	Unconsciousness
	Freezing of extremities
	Core temperature of 95°F or less
	Death

Table 3–3. Effects of poison types

Poison	Targeted Organs or System
Halogenated hydrocarbons	Kidneys
Benzene, nitrates, arsine	Blood
Parathion	Central nervous system
Ammonia, phenols, carbon tetrachloride	Liver

Etiological exposures. An etiological exposure is an exposure to a microorganism that could result in a disease. Most of these diseases are transmitted by blood or body fluids. Simple protective garments provide protection against these diseases. Examples of diseases resulting from etiological exposure include AIDS, tuberculosis, hepatitis, and typhoid.

Radiation. Radiation exposure may result in damage to individuals and future generations. Radiation sickness or poisoning cannot be transmitted to someone else; however, contaminated individuals may carry dangerous material on their bodies, posing a threat to those around them. Radiation may be present in medical centers, nuclear power plants, research facilities, and industrial processes.

The following are protection methods for ionizing radiation:

Alpha particles. Personal protective gear along with SCBA will be sufficient to protect responding personnel from alpha radiation (fig. 3–24).

Fig. 3–24. Turnout gear and SCBA are sufficient to protect responders from alpha radiation.

Beta particles. The use of personal protective gear and SCBA provides only limited protection against beta radiation, which is more powerful than alpha particles. Further shielding can be provided by using heavy plastic, wood, or thin metal. Alpha and beta particles are very dangerous to responders, so decontamination is vital to keeping the scene safe. Also, inhalation or ingestion of alpha and beta particles will make the person radioactive, so care must be taken during decon that a meter is used to ensure the person is clean.

Gamma rays. Gamma rays require dense materials such as lead to shield a person from their effects. Personal protective equipment provides no protection against this type of radiation. Exposure to gamma rays will do damage to the person, but unlike alpha and beta particles, it will not make the victim radioactive.

Neutron. Neutron radiation requires large quantities of water and concrete to shield against its effects, such as is used in the core of a nuclear reactor. Personal protective equipment provides no protection against this type of radiation.

Exposure to radiation can have varying effects on the body, depending on the intensity and duration of the exposure. Large doses can cause extensive cellular damage and result in death, whereas smaller doses can cause cellular damage that increases the risk of cancer. Often the evidence of injury from low or moderate doses may take months or years to appear.

A **chronic exposure** is continuous or intermittent exposure to low levels of radiation over an extended period of time. Effects of chronic exposure to radiation include:

- Genetic effects
- Cancer
- Benign tumors
- Cataracts
- Congenital defects
- Skin changes

Chronic exposure only produces observable effects some time after the initial exposure occurs.

An **acute exposure** is an exposure to a large single dose of radiation or a series of doses over a short period of time. High levels of exposure can result in death within a few hours, days, or weeks. Effects of acute exposure results in the rapid development of radiation sickness and include the following:

- Gastrointestinal disorders
- Hemorrhaging
- Anemia
- Loss of body fluids
- Bacterial infections
- Electrolyte imbalance

Delayed effects of an acute exposure can include the following:

- Cataracts
- Temporary sterility
- Cancer

- Genetic effects

Responders must consider the following when radiation is present: type of radiation, amount absorbed by the body (dose), route of exposure, and length of time exposed.

Radiation exposure reduction methods. The first responder can use time, distance, and shielding to reduce the exposure risk.

- **Time.** As the amount of time spent near a radiation source decreases, the overall exposure is reduced (linear reduction). This results in a proportionate reduction in the dosage received.

- **Distance.** As the distance from a radiation source increases, the amount of exposure decreases. Radiation intensity varies inversely with the square of the distance from the source. As your distance from the source of the radiation doubles, the amount of radiation you are exposed to is decreased by the square of the distance; therefore, doubling your distance decreases your exposure to a fourth of the dose at the original distance. Likewise, tripling your distance reduces the exposure to a ninth of its original strength. For example, increasing your distance by a factor of 2 reduces the intensity of the exposure to ¼ of its original strength (fig. 3–25).

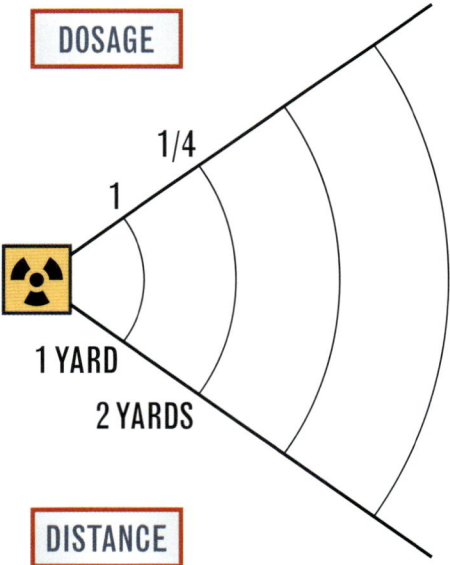

Fig. 3–25. When you double your distance from the radioactive source, you have ¼ of the original exposure.

- **Shielding.** A first responder should use the appropriate type of shielding for the radiation present. Barriers of lead, concrete, or water provide good protection from penetrating radiation such as gamma rays. As discussed previously, the thickness

and the type of materials required to protect from radiation depends on the type of radiation. Keeping substantial objects such as vehicles, dirt mounds, or other obstructions between the source and the responder will help decrease the amount of overall exposure.

Asphyxiants. Simple asphyxiants are typically inert gases that displace the oxygen required for breathing. They dilute the oxygen content of the atmosphere to less than the minimum required for sustaining life. In addition, heavier-than-air gases can pool in low-lying areas such as basements and simply displace the air (fig. 3–26).

Fig. 3–26. Propane, transported by this vehicle, is a heavier-than-air gas.

Chemical asphyxiants are chemicals or substances that prevent the body from using oxygen. They operate in one of three ways: They may tie up the blood's oxygen-carrying hemoglobin and prevent it from carrying the oxygen. They can also remove the hemoglobin from the blood so that there is no longer any transport vehicle for the oxygen. Finally, they may cause other malfunctions in the blood's ability to carry oxygen.

Carcinogen. Carcinogens are materials that are known to cause or are suspected of causing cancer in people.

Convulsants. Convulsants are drugs or chemicals that cause convulsions in exposed individuals. Muscle spasms may start soon after the exposure and may continue for 3 to 30 minutes. Exposure results in dyspnea (difficulty breathing), suffocation, and muscular rigidity. Death can result from asphyxiation or exhaustion. Injury can also be sustained from falling or striking an object on the ground. Examples of convulsants include strychnine, carbamates, organophosphates, and picrotoxin.

Corrosives. Corrosives are chemicals that have destructive effects on materials and burn or destroy living tissue.

When they come in contact with combustibles, they may result in fire or explosion.

Corrosives are divided into two groups: **acids** and **bases**. Contact with acids can result in severe chemical burns to the skin and/or damage to the eyes. Generally, pain is felt upon contact with an acid. Bases react with the body by breaking down fatty tissue and penetrating deep into the body. Contact with bases does not typically result in immediate pain. A greasy or slick feeling on the skin is an indication of contact with a base. This is a result of the fatty tissue breaking down.

The following are symptoms of contact with corrosives:

- Nausea and vomiting

- Burning around the nose, eyes, and mouth

- Difficulty breathing, coughing, or swallowing

- Burning or skin irritation in a localized area

If the potential for exposure to these products exists, first responders should remove themselves from the area and reassess the situation to determine the cause of the symptoms. Any deterioration or discoloration of equipment should alert the first responder to the potential safety issue.

Toxic materials. Some chemicals by their makeup can be extremely lethal at very low levels. Toxic materials can be broken up into two general categories: toxic and highly toxic. The following measurements can determine which category a particular material is assigned to.

Lethal dose (LD50). This is an indication of the lethality of a given substance or type of radiation. The most commonly used lethality indicator is the LD50, the dose at which 50% of subjects will die.

Lethal concentration (LC50). This is the median lethal dose of a toxic substance or radiation. It is the dose required to kill half the members of a tested population. LC50 figures are frequently used as a general indicator of a substance's acute toxicity.

Table 3–4 shows toxicity levels of some common chemical substances. The fire codes International and National Fire Protection Association (NFPA) have provided definitions as to what constitutes a toxic and a highly toxic chemical. The following are the definitions from the 2009 edition of the International Fire Code.

Table 3–4. Toxicity levels of common chemical substances

Chemical	LC$_{50}$ (ppm)	PEL	STEL	IDLH (ppm)
Ammonia	4,000	25 ppm	50 ppm	300
Chlorine	293	Ceiling 1 ppm	—	10
Hydrogen chloride	2,810	Ceiling 5 ppm	—	50
Hydrogen cyanide	—	10 ppm	4.7 ppm	50
Phosgene	5	0.1 ppm	—	2
Sulfur dioxide	2,520	5 ppm	5 ppm	100

Irritant. Irritants cause temporary inflammation of the eyes, skin, or respiratory tract. Symptoms of exposure to irritants include redness, itching, discomfort, and irritated eyes or skin. In addition, coughing or difficulty breathing results from respiratory exposure.

An irritant gives off vapors that target mucous membranes. Although the effect of an irritant is often temporary, it can cause severe inflammation. An irritant may produce permanent damage if exposure is repeated over time.

Sensitizers/allergens. Sensitizers and allergens can cause allergic reactions after repeated exposures. The immune systems of some people believe the material is dangerous and attack it, whereas in other people no reaction takes place. The reaction may be severe or mild and may not occur until several hours or days after the exposure, although most occur within seconds or minutes of exposure. Repeated exposures may cause a rash on the skin or an asthma-like reaction.

Reactions to irritants vary, depending on the means of exposure and individual susceptibility to the material. Some persons may not be affected by the first exposure to the material, but may suffer dangerous effects the next time an exposure occurs. For example, someone stung by a bee for the first time may suffer no ill effects, but upon being stung again may experience severe anaphylaxis or allergic reaction.

Target organ effects. These are signs and symptoms of a chemical exposure that by their nature affect a specific organ or organs rather than the entire body. For example, neurotoxins affect the nervous system, hepatotoxins affect the liver, and nephrotoxins target the kidneys. Upon determining the name of the material and researching its chemical and physical properties, responders should determine the health effects, such as a tendency to affect a target organ, and pass this information to EMS responders.

Psychological. This is mental harm from being exposed to, contaminated by, or even just being in close proximity to an incident. Factors include response to incidents that are devastating to a community, or involve the death of a friend, colleague, or child.

Routes of entry

Individuals may be exposed to hazardous substances through inhalation, ingestion, injection, or absorption. The physical state of matter often determines the mode or route of entry into the body.

Inhalation. Inhalation is the most common route of exposure. Individuals may inhale material such as vapors, gases, fumes, smoke, or aerosols through the nose or mouth. The first responder should protect against this method of exposure by wearing appropriate respiratory protection.

Ingestion. Individuals may ingest materials through the mouth, which commonly occurs when eating, drinking, or smoking contaminated materials. Poor hygiene after handling hazardous materials is the primary cause of this type of exposure. First responders should be completely decontaminated before they are allowed to eat or drink. In addition, no food, drink, or tobacco products should be allowed in the hazmat area.

Injection. An individual may be exposed via injection when materials are taken in through a puncture or needle stick. This mode of exposure is primarily a hazard for emergency medical personnel who handle syringes or IV needles.

Absorption. An individual may absorb materials through the skin or eyes. Many poisons can be absorbed in this manner. The eyes, neck, wrists, groin, hands, underarms, and areas of broken skin are particularly susceptible to exposure by absorption because they are the areas on the body where the skin is the thinnest. Also, any breaks or damage to the skin can allow materials to be absorbed more readily. As a precaution, first responders should always wash their hands prior to touching their face.

Health data terminology

During the research process, responders will encounter the following threshold quantity measurements that indicate the level and types of exposure deemed safe for a given period of time. These values should be compared against the measurements collected on scene to get a benchmark as to how hazardous the level of concentration is. Threshold limit values (TLV) are developed by the American Conference of Government Industrial Hygienists (ACGIH). These values are revised on a regular basis. Threshold limit values can be measured in parts per million (ppm) or in milligrams per cubic meter (mg/m^3) of air.

Threshold limit value/time weighted average (TLV-TWA). The TLV-TWA is the maximum concentration of a substance a person may be exposed to for eight hours a day, 40 hours per week without suffering harmful effects. It is designed primarily for exposures in the workplace. The lower the TLV, the more toxic the material.

Threshold limit value/short term exposure limit (TLV-STEL). TLV-STEL is the maximum exposure time, limited to 15 minutes no more than four times a day without suffering harmful effects. This includes a minimum one-hour rest period between exposures.

Threshold limit value/ceiling level (TLV-C). TLV-C is the maximum concentration that should never be exceeded for any period of time.

Immediately dangerous to life and health (IDLH). IDLH identifies any concentration of a toxic, corrosive, or asphyxiating substance that poses an immediate threat to life. Exposure to these materials can result in irreversible or delayed adverse health effects and may interfere with a person's ability to escape from a dangerous atmosphere. Positive-pressure SCBA is required for entry into an IDLH atmosphere.

Recommended exposure limits (REL). REL is similar to the TLV-TWA and is the occupational exposure limit recommended by the National Institute for Occupational Safety and Health (NIOSH) for a 10-hour work day during a 40-hour work week. These are recommendations based solely on scientific data. It is measured in ppm or mg/m3, just like the TLV.

Permissible exposure limit (PEL). PEL is similar to TLV-TWA and is the maximum allowable exposure in an eight-hour day. This system is adopted by OSHA and is subject to the rule-making and political process. It is measured in ppm or mg/m3, just like the TLV.

HAZARDOUS MATERIALS CONTAINERS

Hazardous materials can be stored and transported in containers of all shapes and sizes. From small laboratory containers to large tank farms holding thousands of gallons of product, operations-level personnel must be familiar with the types of fixed facility and transportation containers found in their district.

Container shapes

The container shape can provide valuable insight into the type of material contained inside. For example, containers with rounded ends and protected valving are most likely carrying a pressurized load, whereas containers with flat ends are typically non-pressurized. Containers with external stiffening rings may indicate a denser substance, such as a corrosive. These simple clues can assist responders very early on in making educated decisions about the contained materials until more specific data can be located.

Bulk packaging and containers (transportation)

There are many mobile tanks and vessels used to transport hazardous materials over roadways, railways, and waterways. The DOT defines bulk packaging as anything that meets the following requirements:

- Liquids: capacity greater than 119 gal (450 L)

- Solids: net weight greater than 882 lb (400 kg) or capacity greater than 119 gal (450 L)

- Compressed gas: water capacity in excess of 1,001 lb (454 kg)

They can be part of the vehicle, such as a cargo tank, or they can be placed on the vehicle, as in the case of intermodals. They are constructed of steel and aluminum; some are insulated and some are not. Some carry cryogenic materials (extreme cold), and some carry molten materials (extremely hot). First responders must have a basic understanding of what these tanks and vessels look like by shape so they can develop an action plan from a distance upon arrival at a transportation incident. Placards as required by DOT may not be visible after a crash or while the tank is burning; however, shapes sometimes tell the story. What follows is a review of the most common types of tanks, vessels, and railcars

you will encounter in your work as a responder. Pay strict attention to the shapes of the containers, which can assist you with identifying the hazard class of the materials.

Over-the-road carriers. Millions of hazardous material shipments are made over highways and streets each year. From gasoline deliveries to the local service station to hydrogen deliveries at manufacturing plants, hazardous materials are all around us on our roadways. Recognition of the different types of container shapes and the materials they carry is important when responding to incidents involving these vehicles.

DOT406/MC306 nonpressurized. These types of tank trucks are used commonly to transport gasoline, fuel oil, kerosene, and other flammable and combustible liquids (fig. 3–27). They are usually constructed of aluminum (and will melt when exposed to fire so the tank does not BLEVE) and run to a normal capacity of 9,000 gal (34,069 L) and an internal pressure of 3–5 psi (21–35 kPa). They are usually compartmented (different grades of gasoline, e.g., regular, premium), have recessed manheads for rollover protection, and are normally egg-shaped or elliptical from the front or rear. Emergency shutoffs are located in the front and rear of the trailer. These tanks may contain gasoline, fuel oil, alcohol, other flammable/combustible liquids, and class B poisons. The main hazards are fire and environmental damage.

Fig. 3–27. DOT 406/MC306 trailer

DOT 407/MC307 low-pressure. These tanks are constructed of steel and may be rubber lined, with a normal capacity of 2,000 to 7,000 gal (7,571 to 26,498 L) and a pressure of under 40 psi (280 kPa) (fig. 3–28). They are also typically double-shelled (tank within a tank) for leak protection and have recessed manheads for rollover protection. Uninsulated tanks have a round cross section with external stiffening rings and a single compartment. Insulated tanks have a horseshoe cross section with one or two compartments. You may find flammable liquids, combustible liquids, poisons, and many other chemicals in these tanks. The main hazards are fire and environmental damage.

Fig. 3–28. DOT 407/MC 307 trailer

DOT 412/MC312 corrosive liquids. These tanks run with pressure up to 55 psi (525 kPa) with a normal capacity of up to 7,000 gal (26,498 L) (fig. 3–29). The tank cross section is round, similar to the MC307 but with a smaller diameter compared to its length. They have numerous rings around the tank shell for strength and stiffening purposes and have built-in rollover and splash protection. The loading area is usually covered with a corrosive-resistant materials. Typically, they will be constructed in one large compartment and contain acids of different varieties. The main hazard is chemical burns.

Fig. 3–29. DOT 412/MC 312 trailer

MC331 high-pressure tanks. These tanks contain gases or liquids under pressure starting at 100 psi (700 kPa) (fig. 3–30). They have capacities ranging from 2,500 to 11,500 gal (43,532 L) and have a bolted manway at the front or rear. The cross section is circular, and it has rounded or dome shaped endheads indicating a pressurized container; the front or rear is round as well. The emergency shutoffs are located in the front and

rear of the trailer. These will also be typically painted white or some other heat-reflective color. These tanks carry propane, liquefied natural gas (LNG), anhydrous ammonia, and other gases that have been liquefied. The main hazard depends on the chemical; however, rapid expansion should be expected.

Fig. 3–30. MC331 trailer

MC338 cryogenic liquids. These tanks contain liquefied gases under pressure starting up to 500 psi (3447 kPa). They have capacities up to 10,000 gal (37,854 L). These are heavily insulated steel tanks that have a round cross section at the front and a squared-off cabinet or compartment at the rear where the valves are (fig. 3–31). Most often they are marked "refrigerated liquid," and you may often see puffs of vapor coming off the tank where the pressure-relief valve is operating. These cryogenic tanks usually carry liquid oxygen, nitrogen, carbon dioxide, and other gases that have been converted to a liquid by refrigeration. The main hazards are frostbite and burns.

Fig. 3–31. MC338 trailer

Compressed gas tube trailer. These units run with pressure between 3,000 and 5,000 psi (21,000 and 35,000 kPa). They are a series of steel tubes or cylinders banded and manifolded together and will typically have an overpressure device for each cylinder (fig. 3–32). The valves are protected with a compartment found at the rear of the trailer. These trailers often carry helium, hydrogen, and other gases under pressure, and can be found at construction or industrial sites. The main hazard depends on the gas; however, high pressures should be expected.

Fig. 3–32. Compressed gas tube trailer

Dry bulk cargo tanks. These tanks are not typically under pressure, but if so, the pressure would be under 22 psi (154 kPa) (fig. 3–33). From the side the tank may look like a series of inverted cones connected together. They use air to assist with loading the dry powders, and the discharge valves are located at the bottom of the hoppers. Dry bulk tanks carry a wide variety of materials, for example bakery flour, plastic pellets, fertilizers, and even slurry-like concrete in bulk. The main hazard depends on the material contained, but toxic materials can also be carried.

Fig. 3–33. Dry bulk trailer

Railcar and tank cars. Railcars are used to transport hazardous materials throughout the country in large quantities. Any department with a railway that passes through its district should be aware of the potential need for a hazmat response. These incidents can range from a leaking container or valve to a large-scale release due to a train derailment. The following are the most common types of rail cars encountered, their characteristics, and the types of materials they carry.

Nonpressure tank cars. Also known as general service cars, these tank cars have a typical capacity of 4,000 to 45,000 gal (15,142 to 170,344 L) and operating pressures between 35 to 100 psi (245 to 700 kPa) (fig. 3–34). Older cars can have an expansion dome with visible fittings. These cars can carry a wide variety of materials including flammables, combustibles, oxidizers, organic peroxides, poisons, irritants, reactive liquids, and other similar materials. The valves and fittings are visible and accessible, and the tank car has flat or nearly flat endheads compared to a pressure tank car. The main hazard depends on the material contained.

Fig. 3–34. Non-pressure tank car

Pressure tank cars. These tank cars can range in capacity from 4,000 to 45,000 gal (15,142 to 170,344 L) and operate at pressures between 100 and 600 psi (700 and 4,200 kPa) (fig. 3–35). They are constructed of steel or aluminum and are of a cylindrical noncompart-mented design. The valves, gauges, and pressure relief valves are located inside a protective housing on the top of the tank.

Flammable liquids and toxic gases are usually found in these types of cars. These tanks have rounded endheads that are more pronounced than nonpressure tank cars. The main hazard depends on the material contained.

Operations Response to Hazardous Materials

HAZMAT

Chapter 3

Fig. 3–35. Pressurized tank car

Cryogenic tank cars. These are heavily insulated steel tanks of a double wall design that contain liquids refrigerated to –150°F (–101°C) or less, at a pressure of 25 psi (175 kPa) or less (fig. 3–36). The fittings on these tank cars are at ground level at either the ends or in the center side of the car in a cabinet. They carry the same cryogenic materials often carried on the MC338 (e.g., liquid oxygen, nitrogen, carbon dioxide, and other gases that have been converted to a liquid by refrigeration).

Fig. 3–36. Cryogenic tank car

Dry hopper cars. These cars vary in configuration and come in three different models: the pneumatically unloaded car, the covered hopper car, and the open-top hopper car (fig. 3–37). The pneumatically unloaded car uses compressed air to assist with unloading what may be dry caustic soda, nitrate fertilizers, plastic pellets, or other fine powders. The open-top hopper is used for heavy coarse material like gravel, coal, and the like. The closed-top hopper transports cement, grains, and similar medium-granulated materials.

Fig. 3–37. Closed top dry hopper rail car

Box cars. With these cars, identification of potential hazardous materials is typically not possible by the car's outer appearance (fig. 3–38). Instead, responders should look for hazard placards. Also, consider that refrigeration and heating systems may pose a hazard.

Fig. 3–38. With box cars, the type of material carried is not readily apparent.

Intermodal containers. An intermodal container consists of a tank container that is placed in a frame (fig. 3–39). This protects the container and allows for easy stacking, lifting, and securing. The frame can be of two types:

- Box type—encloses the tank in a cage
- Beam type—consists of frame structures at the ends of the container

They are called intermodals because they can be used on ships, railways, or highways and are used to transport the same types of products as rail and highway containers. They can also be found in fixed facilities and in nonpressurized or pressurized configurations.

Fig. 3–39. Box type Intermodal container

Nonpressurized intermodals can be found in two types— IM-101 and IM-102—and are the most common type of intermodal used in transportation. They typically have a maximum capacity of up to 6,300 gal (23,848 L) and can be used to carry liquid or solid materials. Despite being listed as nonpressure, some of these intermodals can be pressurized up to 100 psi (700 kPa). They may contain products such as nonflammable liquids, mild corrosives, foods, and other miscellaneous materials, flammable materials, corrosives, and miscellaneous industrial materials.

Pressurized intermodals (Type 5) typically have a maximum capacity of 5,500 gal (20,820 L), and can operate at a pressure of up to 500 psi (3,500 kPa). They may contain products such as liquefied gases, LPG, anhydrous ammonia, and bromine.

Cryogenic intermodals (Type 7) are designed as a tank within a tank separated by a layer of insulation, with the area between the tanks normally containing a vacuum designed to keep the contents cold. The construction is similar to that used in rail and cargo trailers designed for cryogenic materials. These intermodals may contain products such as helium, nitrogen, and oxygen.

Tube module intermodals contain several cylinders 9 to 48 in. (23 to 122 cm) in diameter permanently mounted in an open frame with a compartment at one end that encloses the valving, similar to those trailers seen in rail and highway use. Tube module intermodal containers may contain products such as helium, nitrogen, and oxygen.

Miscellaneous bulk packaging

Intermediate bulk containers (IBC). These containers are commonly found in industry to transport product around an industrial site and for moving bulk chemicals between processing facilities (fig. 3–40). The IBC is usually square in shape and can hold up to 500 gal (1,893 L) of liquid product or an equivalent amount of powders, slurries, or other forms of materials. They can be fabricated of aluminum, stainless steel, or a heavy gauge plastic. The plastic IBCs are often inside a heavy-gauge wire cage for stability and rigidness.

Bulk bags. These have capacities from 500 to 5,000 lb (227 to 2,268 kg). They contain solid materials such as pesticides or fertilizers (fig. 3–41).

Fig. 3–40. Intermediate bulk containers (IBCs) are used to transport materials between facilities.

Fig. 3–41. Bulk bags are used to transport solid materials.

One-ton cylinders. These are capable of transporting one ton of chlorine (fig. 3–42). These cylinders are 3 ft (1 m) in diameter and 8 ft (2.4 m) long with convex or concave heads. They are transported on railcars and trucks and contain liquefied gases such as chlorine, phosgene, and sulfur dioxide.

Fig. 3–42. One-ton cylinders are used to transport chlorine.

Non-bulk packaging or containers

Non-bulk quantities of hazardous materials can be transported using a variety of containers. These can carry many, if not all, of the same materials that the larger containers carry, so the hazards remain similar, but the materials are typically in smaller quantities. The DOT defines non-bulk packaging as containers meeting the following definition:

- Liquids: capacities less than 119 gal (450 L)

- Solids: capacities less than 882 lb (400 kg) or 119 gal (450 L)

- Compressed gas: water capacity of 1,001 lb (454 kg) or less

The following are the most common examples of non-bulk packages.

Bags. Many dry materials come in bags for transport and use in industrial fixed facilities (fig. 3–43). The contents could be pesticides, oxidizers, organic peroxides, and the like. Sometimes they are simple powders

or granule materials. The bags can be made of woven materials, plastic, or paper and are sealed by heat, glue, ties, or stitching. Bags are normally transported and stored on pallets.

Fig. 3–43. Bags can contain various types of hazardous materials.

Boxes. These may be fiberboard (cardboard) or wood (fig. 3–44). They may contain almost any type of hazardous material such as liquids, solids, radioactive materials, and so on.

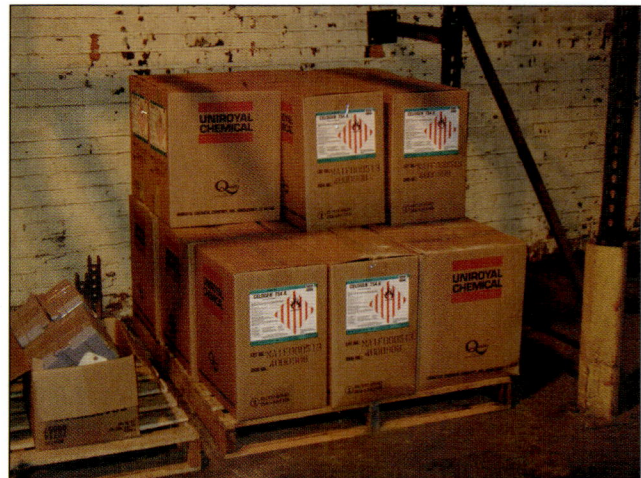

Fig. 3–44. Boxes can be used to transport virtually any type of hazardous materials.

Small containers. Common small containers of chemicals range in size from a few milliliters to 5-gal (19-L) cans (fig. 3–45). These types of containers are most often found in laboratories and research facilities where smaller containers are easy to store and easy to handle by the research staff. They are used for flammable liquids (solvents), corrosives, acids, and other lab materials needed in small quantities. These smaller containers can be made of metal, plastic, or glass.

Fig. 3–45. Small containers are often found in labs.

Carboys. These containers can be made of glass, metal, or plastic (fig. 3–46). Glass carboys can be found in a fiberglass or wood box for protection from breakage. Carboys are designed to hold liquids and can have a typical capacity of 5 to 15 gal (19 to 57 L). They may contain water, poisons, pesticides, herbicides, caustics, acids, or flammable/combustible solvents. The type of material the carboy is constructed of depends on the material carried.

Fig. 3–46. Carboys are designed to carry various liquids, usually strong acids.

Dewar flasks. Liquid substances that must be kept cold can be shipped or stored in a Dewar flask (fig. 3–47). It is a double-walled container with a vacuum between the layers to provide thermal insulation, essentially a tank within a tank. One of the more common uses in a laboratory is to hold liquid nitrogen.

Fig. 3–47. Dewar flasks carry cryogenic liquids.

Drums. Drums are used to store and transport the entire gamut of hazardous materials (with a few exceptions), including but not limited to flammable liquids, combustible liquids, corrosives, water-reactive chemicals, oxidizers, acids, and the like (fig. 3–48). They can be constructed of metal, plastic, or a fiberboard material. It should be noted that not all drums are 55 gal (208 L) in size, although that's the most common. The 33 gal (125 L) drum can be found in fixed facilities as well. Also used to pack powders and liquids are fiber drums in various sizes, which look like cardboard but are treated and constructed of a tougher material than normal corrugated cardboard used in a standard delivery box. Getting it wet, however, may lead to container failure. Common industry practice when using fiber drums is to line them with heavy plastic bags or inserts.

Operations Response to Hazardous Materials

HAZMAT

Chapter 3

Fig. 3–48. Drums come in various types and sizes and are used to carry just about any type of material.

Compressed gas cylinders. These cylinders are as common as storage tanks and drums in fixed facilities, and in certain industries they are more common (fig. 3–49). The semiconductor, electronics, pharmaceutical, and medical industries are some examples of industries in which compressed gases are used on a daily basis. Cylinders can range in size from a few pounds to several thousand gallons. Common household use of cylinders can range from a 20-lb (76-lb) gas grill propane cylinder to a 250-lb (946-lb) cylinder used to heat a home. Cylinders can hold flammable gases, toxic materials, poisons, corrosives, or mixtures of these at pressures between 40 and 900 psi (280 and 6,300 kPa). Cylinders are outfitted with an operation wheel for opening and closing and, with the exception of those storing poisons and similar toxins, have some type of relief plug or frangible disk to prevent overpressuring. This will operate when heated to relieve the pressure on the cylinder when impinged upon by fire or other heat source. The relief should allow the pressure to blow off to the atmosphere, rather than causing a catastrophic failure of the cylinders, which are constructed of rolled steel. This prevents flying shrapnel, which could cause injury or death.

Fig. 3–49. Compressed gas cylinders come in many sizes.

Radioactive packaging

There are several types of packages specially designed to transport radioactive materials. Each of these packages has different requirements depending on the type of radioactive materials being transported.

Excepted packaging is used to transport materials with extremely low levels of radiation (fig. 3–50). It is used for limited quantities of materials, instruments, and items such as smoke detectors, and consists of fiberboard boxes and wooden or steel crates. Packages and shipping papers do not indicate a hazardous material.

Fig. 3–50. Excepted packaging is used for materials with extremely low levels of radiation. (Courtesy of DOE)

Industrial packaging is used to transport materials that present a limited hazard (fig. 3–51). They may contain contaminated equipment or radioactive waste encapsulated in concrete. There are three different categories of industrial packaging based on strength ratings.

Fig. 3–51. Industrial packaging is used for materials that present a limited radiation hazard. (Courtesy of DOE)

Type A packaging is used to transport low-level radioactive wastes and radiopharmaceuticals with a higher concentration than is allowed in excepted and industrial packaging (fig. 3–52). Generally, type A packages have an inner containment vessel of glass, plastic, or metal and packing material of polyethylene, rubber, or vermiculite.

Fig. 3–52. Type A packaging is used to transport low-level radioactive wastes. (Courtesy of DOE)

Type B packaging is used to transport spent fuel, high-level radioactive wastes, and high-concentration radioisotopes (fig. 3–53). They range in size from a small radiography cameras to 55-gal (208-L) drums to heavily shielded steel casks that weigh 98 tons or more.

Fig. 3–53. Type B packaging is used to transport high-level radioactive wastes. (Courtesy of DOE)

Type C packaging was created for the transport of more highly radioactive material by air and is designed to withstand severe air crashes. It has the same basic shape as Type B package. This type of packaging is rarely used.

Fixed facility containers

Much like transportation containers, fixed facility containers come in a variety of shapes and sizes and are designed for the materials they hold. Many of the same concepts regarding tank design presented in the previous section will apply to these containers as well. Fixed facility storage is found in low or nonpressure, pressurized, and other types.

Low pressure or non-pressure bulk storage. Low pressure or nonpressure fixed facility storage includes the following types.

Cone roof. The roof is an inverted cone welded to the tank side with weak seams so that the roof could easily blow off if the tank explodes or overpressurizes (fig. 3–54). This type of facility could contain any combination of flammable or combustible liquids and may have a fixed foam system.

Fig. 3–54. Cone roof tank

Horizontal tanks. These tanks sit on legs, usually a steel frame on a cement pad (fig. 3–55). Horizontal tanks can contain any combination of flammable or combustible liquids and may have some form of fixed protection (e.g., deluge or foam spray system).

Fig. 3–55. Horizontal tank

Floating roof tanks. The roof is set on floating pontoons that float up and down on the surface of the product contained therein (fig. 3–56). These are mostly used for gasoline storage where the floating roof eliminates vapor space, lessening the hazard.

Fig. 3–56. Floating roof tank

Covered top floating roof tanks. This type of tank looks similar to the cone roof tank, except that there are eye vents around the top rim of the tank (fig. 3–57). Like the floating roof tank, there is an internal pan that floats on the surface of the product. These tanks are mostly used for bulk storage of gasoline. A similar design to this type of tank consists of a roof shaped like a geodesic dome, also used for bulk gasoline storage.

Lifter roof tanks. The roof slides up and down on tracks, and when the contents pressurize, the roof lifts slightly to relieve the pressure. This tank is used for gasoline and other high vapor-pressure flammable liquids.

Underground storage tanks (non-pressure). These are found mostly in gas stations where the tanks are buried below grade, and the materials are pumped to the surface for dispensing.

Fig. 3–57. Covered tank with internal floating pan

Dome top/dome roof. These are low-pressure tanks for flammable and combustible liquids stored up to 15 psi (105 kPa) (fig. 3–58).

Fig. 3–58. Dome top/dome roof tank

Pressurized fixed facility storage. Storage types for pressurized fixed facilities include the following.

Spherical pressure vessels. High-pressure storage for up to 600,000 gal (2.3 million L), these vessels are used mostly for liquefied petroleum gas (LPG) and vinyl chloride (fig. 3–59). It is a single shell tank of noninsulated design with operating pressures ranging from 100 to 500 psi (700 to 3,500 kPa). Like water storage tanks, these tanks are often sited on large steel support legs.

Fig. 3–59. Spherical pressure tank

Horizontal pressure tanks. These are high-pressure storage tanks with capacities from 500 to 40,000 gal (1,893 to 151,416 L) of product at 100 to 500 psi (700 to 3,500 kPa) (fig. 3–60). Its rounded ends indicated a pressurized container. They are typically of a single shell, noninsulated design and are usually painted white or a highly reflective color to reflect light and heat. These tanks are used most often for propane, ammonia, butane, ethane, compressed natural gas (CNG), and similar materials.

Fig. 3–60. Horizontal pressure tank

Other types. Additional types of fixed facility storage include the following.

Cryogenic liquid tanks. Pressures vary from product to product and can range up to 250 psi (1,750 kPa) with a typical capacity of up to 1,000 gal (3,785 L) at most fixed facilities (fig. 3–61). As with cryogenic transport containers, they are insulated and vacuum jacketed. They can be found at hospitals, gas processing plants, and industrial plants. They are used for carbon dioxide, liquid oxygen, liquid nitrogen, and similar materials.

Fig. 3–61. Cryogenic liquid tank

Mounded above-ground tanks. In older terminals, usually fuel oil depots and the like, tanks that are half in-ground and half above-ground may be found, mounded over with dirt and grass or gravel. These tanks are connected to a fuel oil truck loading rack and are protected with a bulk foam system.

ESTIMATING POTENTIAL HARM

The focus of emergency response personnel at the operations level is to make general estimates regarding the area that will be affected by the release by using the hazard and response information collected. Responders have a number of tools at their disposal to estimate the potential harm that a hazmat release will have on the surrounding area. When these estimates are made, evacuations can begin, and defensive operations can be safely initiated. If the situation warrants it, the response of a technician-level team or other high-level resources facilitates more detailed estimates of potential harm.

Determining the size of an endangered area

One of the easiest and most readily available tools for estimating the size of an endangered area is the DOT ERG. As discussed previously, both the orange emergency response guide section and the green tables of isolation and protective action distances contain information on isolation and evacuation distances. Although this may seem redundant, this system accounts for the fact that some materials present a toxic-by-inhalation hazard (TIH) whereas others do not. The distances for TIH materials can be found in the green section, whereas the other ones are determined by looking in the orange guide section. Use the flow chart in (fig. 3–62) to choose the appropriate section of the ERG to use in determining the size of the endangered area.

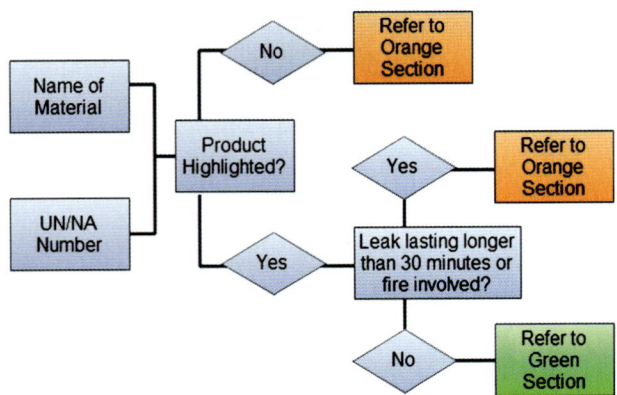

Fig. 3–62. Determining the size of the endangered area using the DOT ERG

If the only method of identification visible is a placard or a container shape, use the white section of the ERG to reference the appropriate guide number in the orange section (fig. 3–63).

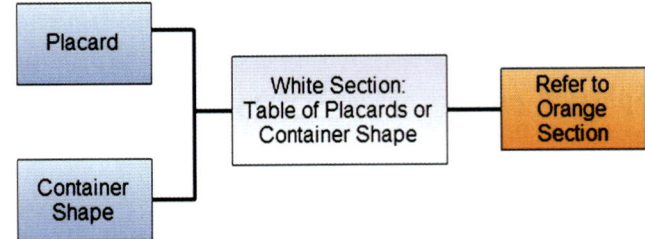

Fig. 3–63. Determining the size of the area using the container shape or placard and the DOT ERG

To control the endangered area, the hazmat responder should:

- Isolate the area of the spill using the initial isolation distances listed in the ERG.

- Remove people from the initial isolation area.

- Deny entry to the isolation area by unauthorized persons.

- If required, evacuate or protect-in-place those people found within the protective action zone.

Initial isolation and protective action distances

Determining the initial isolation zone. The first responder should reference the proper section of the ERG to determine the initial isolation zone. The public safety section in the orange guide should be used to determine the isolation zone for materials that are *not* highlighted. Evacuation distances are listed for materials involved in fires as well as spills. Materials that are highlighted use the initial isolation recommendations given in the green section. The initial isolation zone is established by creating a circle around the spill using the minimum isolation distance, as shown in figure 3–64.

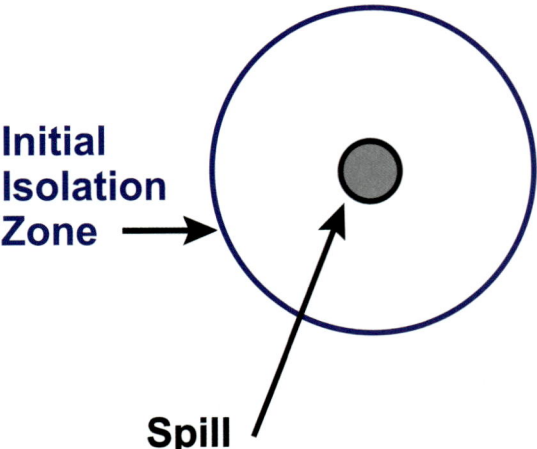

Fig. 3–64. Initial isolation zone layout

Determining protective action distances. The protective action distance is the area that is adjacent to and downwind of the initial isolation zone and in imminent danger of becoming contaminated within 30 minutes of the material's release (fig. 3–65). Responders should consider evacuating or protect-in-place the people found in this area. Isolation distances for highlighted materials can be found in Table 1 of the green section of the ERG. Materials that are *not* highlighted do not have specific protective action distances given. The ERG suggests that you increase the initial isolation zone distance given in the orange guide in the downwind direction as necessary.

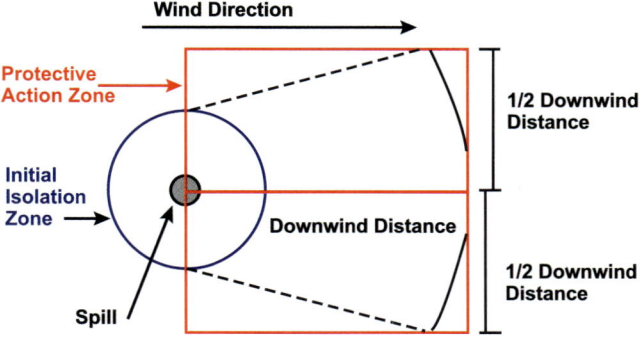

Fig. 3–65. Protective action distances layout

The protective action zone is shaped like a square and provides downwind protection.

Also, CHEMTREC and/or the manufacturer or shipper can provide additional information to supplement what is gathered from the DOT ERG. Other methods available to response personnel, depending on their level of training and equipment available, are computer plume modeling and metering capabilities to determine the extent of the release.

Estimating the potential exposures

After the size of the endangered area has been determined, response personnel must look at what types of exposures are in that area. In order of importance, exposures consist of people, the environment, and property. Responders should consider the potential extent of physical harm that can be anticipated in the area and what would be accomplished by taking action or not taking action before making any determination on an appropriate response. Factors in that decision include the number of exposures, to what extent they are exposed, and the concentration of the released material in the area of the exposures. Also, factors such as whether it is a rural or urban area, time of day, and the general occupancy in the area will all play a role in that estimation. Additionally, the size of the container, the size of the breach, as

well as wind direction and speed will also be important factors to consider. Responders should be keenly aware of the scope of potential exposures that exist within the danger area and be sure to request adequate response resources that will support minimizing the effect on those exposures. This includes additional evacuation assistance and technician-level responses to potentially reduce the amount of product released.

Determining the concentration of a material

The two primary methods of determining the concentration of a chemical in a given area is the use of appropriate monitoring or metering equipment and computer-based dispersion model programs. These resources are not typically available to operations-level responders, so access to this equipment and personnel trained to use it should be arranged prior to an incident. Responders should be sure to train with these additional resources prior to ensuring that their capabilities are understood.

Atmospheric monitoring. Atmospheric monitors are used by firefighters to detect and measure the concentration of airborne health and fire hazards that may be present following a chemical release, a structure fire, or a confined space release. These hazards exist both as gases and vapors and may be produced by liquids or solids. Remember, liquids and vapors are readily inhaled, and those that are flammable are easily ignited when mixed in the appropriate concentration with air.

Most atmospheric monitors provide responders with a numerical reading that measures the concentration of products in the atmosphere. These numbers are then compared to standards that define concentration levels considered safe. Many atmospheric monitors are already programmed with these levels and include both audible and visual alarms that sound when unsafe concentrations are reached.

The devices used most often by responders have four separate sensors and are, therefore, usually referred to as **4-gas meters**. A typical atmospheric monitor includes a small electric pump that draws in a sample of the atmosphere. Today, most fire department monitors have four sensors: a lower explosive limit (LEL) sensor that detects flammable atmospheres, an oxygen sensor to monitor oxygen levels, a carbon monoxide sensor to determine carbon monoxide levels in parts per million, and a hydrogen sulfide sensor to monitor levels in parts per million. Each of these sensors is designed to detect and measure only one chemical. Figure 3–66 shows a typical 4-gas meter.

Fig. 3–66. A typical 4-gas meter

For most toxic atmospheres, measurements are usually read as **parts per million (ppm)**. Oxygen content is measured as its percent concentration in air, normally 20.9%. Monitoring for oxygen is important, and the alarm values of 19.5% and 23.5% are set by OSHA regulation, but the reality is that responders can be in danger before the alarms sounds. An oxygen drop to 20.8% from 20.9% can be a dangerous situation depending on the contaminant that has caused the drop. It may be only 0.1%, but that is a significant change (1% of air = 10,000 ppm, so a drop on the O_2 reading of 0.1% = 1,000 ppm). A drop to 19.9% would not result in an alarm, but it indicates considerable contaminants in the air. The concentration of flammable gases and vapors is typically measured as a percentage of the material's LEL, which is the lowest concentration when mixed with air that allows the material to burn once ignited. There are standard set points for the meter's alarm to sound when it detects unsafe levels of the monitored materials. Figure 3–67 provides alarm set points for an atmospheric monitor.

Gas	Ceiling	TWA	STEL
CO	35 ppm	35 ppm	100 ppm
H_2S	10 ppm	10 ppm	15 ppm
SO_2	2 ppm	2 ppm	5 ppm
O_2	Low: 19.5% High: 23.5%		
LEL	10% LEL		

Fig. 3–67. An example of alarm set points for an atmospheric monitor

Atmospheric monitors do not *detect* immediately when exposed to an atmosphere. The manufacturers determine in advance the **response time** of the device, which is the maximum amount of time it may take to obtain a reading. The use of probes or flexible tubing can increase this time. Most manufacturers list the response time for each sensor in the manual. To use the device appropriately, make sure that the response time is understood.

The ability of an atmospheric monitor to provide accurate readings must be checked on a regularly scheduled basis. This check is referred to as **calibration** and is performed by exposing the atmospheric monitor to a predetermined concentration of gases. If the readings obtained are inaccurate, adjustments or repairs may be necessary. Calibration must be done usually every 30 days or based on the manufacturer's recommendation, and only by those with the proper training to conduct these tests safely.

Remember that the type of gas used to calibrate an atmospheric monitor influences readings that measure the concentration of flammable gases and vapors. Although there are many different flammable gases and vapors responders could encounter during a salvage and overhaul or a hazmat emergency, the atmospheric monitor is only *calibrated* to read *one* type of flammable gas. Readings obtained when attempting to detect and measure the concentration of other types of flammable gases may be higher or lower than their actual concentration. Such readings must be adjusted using a **relative response factor** provided by the manufacturer of the detection device.

For example, if an atmospheric monitor is calibrated using methane gas, readings when measuring the concentration of this gas are correct. However, when attempting to measure the concentration of another gas such as propane, the manufacturer of the atmospheric monitor may indicate that the reading must be *corrected* by multiplying it by 1.5. It is important to remember that each gas or vapor has its own response factor that is specific to each type of device. Always refer to the manufacturer's manual for your instrument.

When using atmospheric monitors, responders must protect themselves from the hazardous atmosphere they are in. Products of combustion from structural fires or gases and vapors from a hazmat incident are not always visible. Precautions for toxic and flammable atmospheres include the use of structural firefighter's protective clothing and SCBA. Some toxic materials may be absorbed through the skin despite wearing structural firefighter's protective clothing, as these garments are not airtight. Remember, readings that exceed the safe limit for the concentration of flammable gases and vapors demand the immediate evacuation of personnel and equipment to a safe area.

Responders using atmospheric monitors must take into consideration the physical state (gas, liquid, or solid) and chemical properties such as flammability, corrosiveness, toxicity, and high energy of chemicals. For example, some gases and vapors, such as methane, are lighter than air, whereas others, such as propane and gasoline vapors, are heavier than air. Because of a product's vapor density, a responder holding an atmospheric monitor at waist level may not obtain accurate readings.

Many atmospheric monitors have adjuncts that help sample the atmosphere. These adjuncts may include probes or flexible tubing that can be inserted into openings above or below grade. When using these probes, responders must be careful not to place the probe or tubing into liquids or unknown areas without a filter provided by the manufacturer. Figure 3–68 shows a typical probe used with atmospheric monitors.

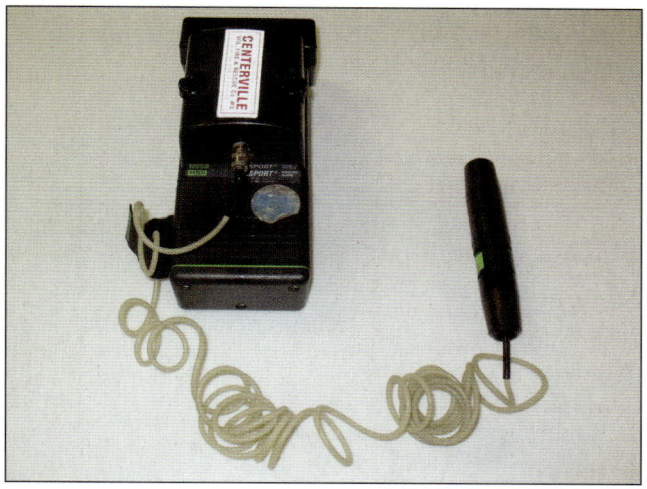

Fig. 3–68. Typical probe used with atmospheric monitor

Telephone resources

Many hazardous materials experts may not be located in your area, but they are only a phone call away. There are several 24-hour hazardous materials emergency resources available to the first responder.

Chemical Transportation Emergency Center (CHEMTREC).
CHEMTREC is a 24/7/365 emergency call center that provides immediate information and assistance to anyone involved in a chemical or hazardous material incident. It can be contacted by calling 1-800-424-9300.

Today, CHEMTREC is the most widely used resource for emergency responders dealing with hazardous materials incidents. The service is free to all responders and anyone who deals with hazardous material emergen-

cies. CHEMTREC is staffed by trained and knowledgeable personnel who have access to over 3 million MSDSs. CHEMTREC staff can fax or email an MSDS to you on scene.

CHEMTREC's emergency service specialists (ESS) can provide immediate guidance and links to technical expertise, as well as medical and toxicological advice and information through MSDSs and other resources. Also, information about the contents of tank cars, trucks, or other containers is routinely provided during emergencies through carrier contacts, shipping documents, and other systems and resources available to CHEMTREC ESS.

According to law, companies that ship hazardous materials must provide a 24-hour emergency contact number. CHEMTREC provides this service to its members. CHEMTREC then maintains a database of the company's MSDSs as well as emergency points of contact. Responders today can call CHEMTREC at any time, even when the company whose product has been released is not registered with them. CHEMTREC emergency service specialists will still try to assist the responder with information about the chemicals and the shipper. For additional information, visit http://www.chemtrec.com. There are versions of CHEMTREC that exist in Canada and Mexico that provide a similar service. In Canada the service is known as CANUTEC (Canadian Transport Emergency Center), in Mexico it is called SETIQ (Stema de Emergencia para la Transportacion de la Industria Quimica). When contacting any of these agencies, it is important for the responder to provide the following information. This information should also be provided to local, state, and federal resources when requesting their response to a hazardous materials incident:

- Full name and title of caller
- Callback telephone number
- Location and description of incident
- Time of incident and weather conditions
- What actions have been taken thus far
- Type or description and number of containers or packages involved
- Product info from shipping documents
- Name of shipper
- Name of carrier
- UN, NA, and/or STCC number of the product

- Trade name or chemical/DOT shipping name
- Shipper and point of origin
- Consignee and destination
- Amount of material involved
- Identification numbers on the container
- Any fatalities, injuries, or exposures
- Specific information requested or type of information that is needed as a priority
- What resources are on scene or en route

Response personnel should be aware of the limitations of the CHEMTREC system because not all chemical manufacturers provide information to these organizations. The relevance of the information provided by CHEMTREC depends upon the information given to them by the caller. The first responder must understand that the response procedure information given is general in nature.

National Response Center. The National Response Center, located in Washington D.C. and staffed by the U.S. Coast Guard, is the single point of contact for reporting spills of oil, hazardous materials, and terrorism incidents. The National Response Center is also staffed 24/7 and can be contacted at 1-800-424-8802.

Many hazardous materials have a **reportable quantity (RQ)**. If the amount of material spilled is more than the RQ, the spiller is required to notify the National Response Center. The National Response Center then notifies the U.S. Coast Guard and/or the EPA, depending on the spill location, as well as state agencies and other government entities. This is how the state and federal governments learn about some of the spills that occur in your area.

Centers for Disease Control. If a victim is involved in your incident, you need to quickly find out what care must be administered. Your local poison control center may be this first call. If your local center has limited expertise in hazardous materials, you can also contact the Centers for Disease Control and Prevention, which has a division known as the Agency for Toxic Substances and Disease Registry (ATSDR), composed of a team of toxicologists who specialize in environmental health and hazardous materials. ATSDR can be reached 24 hours a day at 404-498-0120. If a contaminated victim is transported for treatment, responding medical personnel may want to use this resource to gain additional information regarding treatment and decontamination.

Computer resources

Many computer resources are available to the emergency responder responding to a hazardous materials incident. Two of the most common resources are discussed in the following sections.

WISER. Wireless Information System for Emergency Responders (WISER) is produced and distributed by the National Library of Medicine (fig. 3–69). WISER is a free download that can be accessed at http://wiser.nlm.nih.gov. WebWISER is also available for mobile use and is also available as an app from any app store. A responder can subscribe to WISER and receive information on updates and online training exercises. WISER is one of the most commonly used tools by hazardous materials responders. WISER contains information on more than 4,700 hazardous substances. Furthermore, WISER also has extensive search capabilities, which can be especially helpful in an incident where limited information is available.

Fig. 3–69. Wireless Information System for Emergency Responders (WISER)

CAMEO. **Computer-Aided Modeling for Emergency Operations (CAMEO)** was developed by the National Oceanic and Atmospheric Administration (NOAA) and the U.S. EPA (fig. 3–70). CAMEO consists of three modules that can be useful to the emergency responder. CAMEO is the portion that users can load facility information and will integrate Tier II reports from eplan. CAMEO Chemicals contains a chemical database with more than 6,000 chemicals. The user can access the response information data sheets (RIDS) that contain similar information to that on an MSDS. CAMEO also contains a chemical reactivity worksheet that predicts possible outcomes when chemicals are accidentally mixed together, such as at a transportation incident. CAMEO is available for free download at www.noaa. gov. It takes some practice to master, but CAMEO can provide valuable information for planning and response.

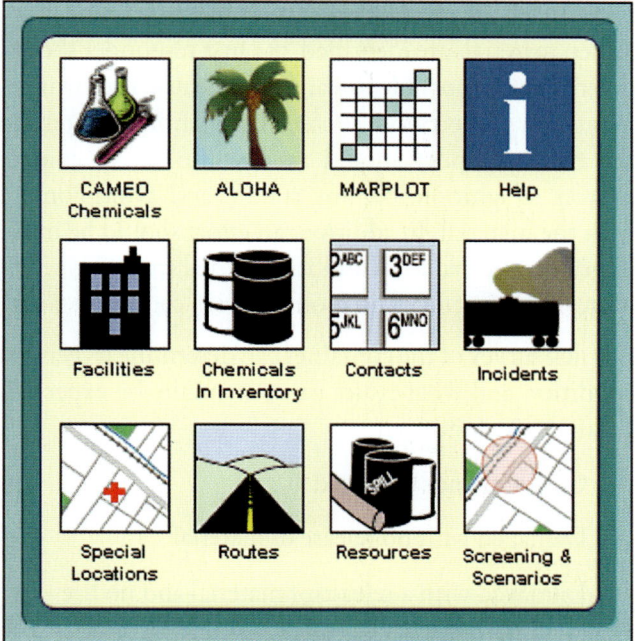

Fig. 3–70. CAMEO provides a comprehensive computer-based resource for responders.

Another module of CAMEO is A real Locations of Hazardous Atmospheres (ALOHA). A plume model that can estimate hazard zones for a chemical that becomes airborne, ALOHA requires the user to input information about the incident such as type of spill, weather conditions, and surrounding environment.

The third module of CAMEO is a mapping tool known as the Mapping Application for Response, Planning, and Local Operational Tasks (MARPLOT). MARPLOT allows the user to map the incident directly on a location where the incident is occurring. The user can then use ALOHA for a visual estimate of the plume of the released hazardous materials.

Decontamination

Decontamination is a process of removing or neutralizing contaminants that have accumulated on personnel, equipment, and the environment. Decontamination protects responders from agents that may contaminate and permeate the protective clothing, SCBA, tools,

apparatus, and other equipment used at the incident. It protects all on-scene responders by minimizing the transfer of harmful materials to clean areas and helps prevent the mixing of incompatible chemicals. It also protects the community by preventing uncontrolled transportation of contaminants from the incident. Decontamination operations can be hindered by the lack of a positive identification of the material, weather conditions, topography, and availability of equipment and trained personnel. Operations level responders will focus on conducting Gross Decontamination and Emergency Decontamination. General information on decontamination and detailed information on gross and emergency decontamination can be found in chapter 4 Mission Specific Decontamination section.

Personal protective equipment

Personal protective equipment is an important component in any hazmat responder's inventory. More important, however, is the knowledge of how to pick the appropriate level of protection for the incident. Failure to understand how to interpret hazard and response information and use it to choose the correct PPE can result in significant injury to response personnel. Responders should also be trained in the use of any specialized hazmat/WMD PPE issued by the authority having jurisdiction (AHJ). The two major categories addressed here are respiratory protection and personal protective clothing. Operations-level responders must be specifically trained to use chemical protective equipment (CPE) and in the operation of technical decontamination before they are permitted to wear CPE at any hazmat/WMD incident. Also, they must know the differences between the various levels of protection and understand how to choose the appropriate level of PPE for a given material and how the available level of protection affects the options for various operations. If there is a lack of appropriately trained personnel and/ or equipment, a request for personnel with the proper level of training and equipment should be transmitted immediately.

Typically, operations-level responders performing defensive actions will only have structural firefighting protective clothing and SCBA available to them. If the authority having jurisdiction (AHJ) can decide to provide chemical protective clothing to their operations-level responders. This would require mission specific training in PPE and decontamination.

Types of response actions

The three types of response actions include nonintervention, defensive, and offensive.

Nonintervention response. Nonintervention responses are implemented when the incident is beyond the capabilities of the responders, an imminent explosion is probable, there is a potential for a massive release, or the LERP or facility plan dictates this approach as determined by a site preplan.

The first responder should take the following actions when a nonintervention response is appropriate:

- Relocate responders and other personnel to a safe area.
- Relay conditions to dispatch and request additional resources.
- Establish scene control and an incident management system.
- Evacuate or protect-in-place as necessary.

Defensive responses. Defensive responses are implemented when personnel have adequate training and equipment to confine the product or the LERP or facility plan dictates this approach as determined by a site pre-plan of the hazards (fig. 3–71). This is the primary response action for the operational-level responder. The first responder should take the following actions when a defensive response is appropriate:

- Relay incident conditions to dispatch center.
- Set up an incident management system and control zones.
- Evacuate or protect-in-place as necessary.
- Use defensive operations to control and confine the product.
- Request additional resources as required.

Fig. 3–71. Defensive operations can take place when responders have adequate training and equipment.

Offensive responses. Offensive response is limited to technician-level responders because first responders at the operational level are not trained to initiate offensive operations. The exception to this rule involves materials such as gasoline and propane when the operations-level responder has received specific training on handling the hazard.

Strategic considerations

When materials are identified, the first responder should determine the possible threat to life, health, the environment, and property. In the case of unidentified materials, the first responder should assume the worst, control sources of ignition, confine runoff, and avoid contact with the material. In addition, an effort should be made to protect the material from any external stressors such as heat, shock, or contamination that may cause it to react.

A quick survey of the container can determine its general condition and what issues can potentially be expected. Examples include:

- Container not damaged
- Damaged with no release of material
- Damaged with a release of material and no fire
- Damaged with a release of material and fire

Strategic priorities

It is impossible to determine the overall objectives of the response without first looking at the strategic priorities that will define the tactics used to handle the response. As mentioned previously, the fire service's incident priorities in any emergency are life safety, incident stabilization, and property conservation. The following, in order of importance, is a more specific breakdown of these priorities when responding to incidents involving hazardous materials:

- Rescue operations
- Protect exposures (people and property)
- Fire control
- Spill control
- Leak control
- Recovery

Rescue operations. Rescue operations should be the first priority at any incident (fig. 3–72). The first responder should analyze the risk versus benefit and determine if their PPE will provide adequate protection to effect a rescue.

Fig. 3–72. Rescue operations should be a first priority at any incident.

In multiple-victim situations, the first responder should rescue victims who can be easily saved first. The first responder must also be aware of the potential for rapid deterioration of conditions. A minimum of emergency decontamination must be established prior to rescue attempts.

During rescue operations, consider the following:

- What is the nature of the material involved and the severity of the incident?

- Is appropriate PPE available?

- How many victims are there? What are their conditions?

- How long will rescue operations take?

- Is the necessary equipment available to complete the rescue?

Protection of exposures. When attempting to protect an exposure, the first responder should determine the number of exposures involved in the incident and decide how many have already been lost. The first responder must also determine the best manner of protecting people, the environment, and property from further damage. Depending on the material released and its chemical and physical properties, it may be necessary to request additional resources to mitigate the release and/or isolate people in the area involved.

When protecting people, the most important consideration is life and health. The first responder can evacuate or protect-in-place to protect potential victims from the hazardous material.

Environmental concerns such as the atmosphere, surface water, wildlife, water table, and surrounding land should also be protected from exposure. Note that it may take years to fully realize the effect of an incident.

To protect property from exposure, the first responder should consider the extent of contamination of exposed items. Dilution or diversion of hazardous materials may be the best option. Do not risk life or the environment to save property. It may be necessary to write it off if the danger is too great.

Fire control. In most cases fires involving hazardous materials should be left to burn until responders can determine if extinguishment is the correct option. However, first responders may have to extinguish fire in order to conduct rescue operations. This tactical priority may be moved up in its importance if necessary to rescue people or to protect them from exposure. Careful evaluation is necessary prior to committing emergency personnel to extinguishing fire, as it may worsen the situation. Allowing the material to burn may be a viable alternative (fig. 3–73).

Fig. 3–73. In situations where extinguishing the fire can make things worse, allowing the material to burn is a viable alternative.

Spill control confinement. This is a defensive action in which the flow of the hazardous material is controlled (fig. 3–74). Operations level responders often use confinement techniques to control the spread of the material. These techniques include damning, diking, diverting, vapor suppression and dispersion. Responders should make every effort to stay out of the product and confinement techniques should occur outside of the hazard zone. Detailed information about confinement techniques can be found in chapter 4 Mission Specific Product Control section.

Fig. 3–74. Confinement is a defensive action in which the flow of the hazardous material is controlled by Operations Level responders.

Leak control containment. This is an offensive action that prevents additional releases from a container (fig. 3–75). It is not considered a first-responder skill and is performed by technical-level responders.

The following are factors affecting containment:

- The condition of the container
- Chemical/physical properties of the material
- The material's rate of release

Fig. 3–75. Containment is an offensive action that prevents additional release from the container.

Public protection actions

Evacuation. Evacuations are needed if the situation is such that the at-risk population is at a greater risk by remaining near the incident site. If there is enough time, evacuation is the best protective action. First responders should begin by evacuating people nearby and those outdoors in direct view of the scene. Evacuees should be sent to a definite place by a specific route, far enough away so they will not have to be moved again if the wind shifts. When conducting an evacuation, responders should instruct evacuees where to go, move evacuees to a safe location, provide evacuees with basic needs, and keep them informed of the incident's status (fig. 3–76).

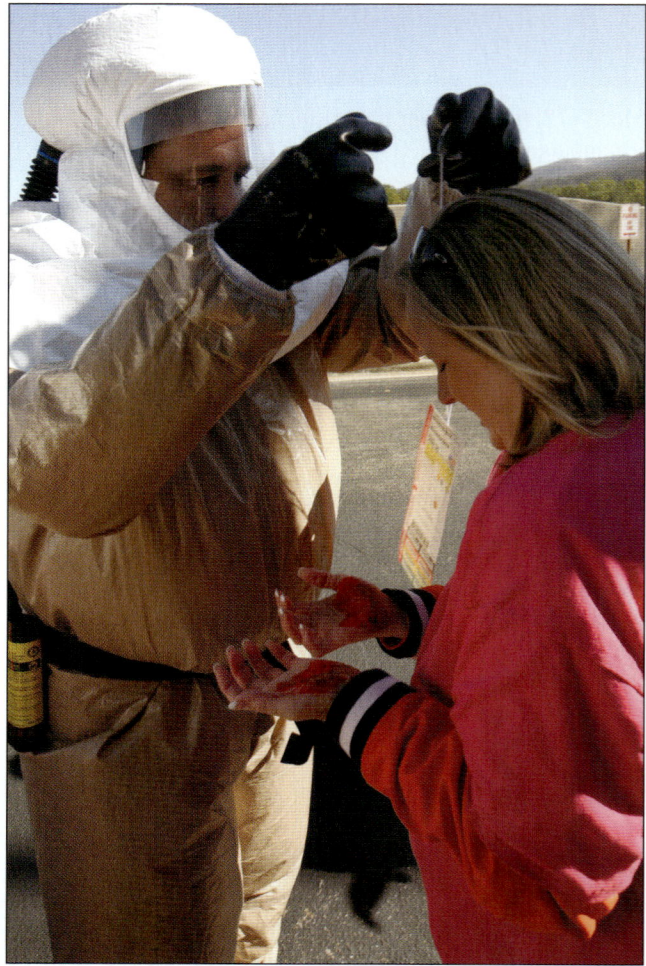

Fig. 3–76. Responders need to advise evacuees on the best place to relocate.

Evacuations should be conducted when the incident involves uncontrollable leaks, explosive materials, or unknown gas leaks from large-capacity containers or basically when the release is going to long term. The following are factors to consider when making the determination to evacuate:

- The degree and severity posed by the hazard based on the risk analysis developed by response agencies and the LEPC

- The number of persons or the size of the population area that will be affected by the hazard

- The amount of resources (fire/police/EMS/private or public buses/mass transit) needed to effectively evacuate the population at risk

- The available resources needed to notify and provide instructions to the public, including radio and television media, mobile public address systems, reverse 9-1-1, door-to-door contact, cell phone text messaging, and recent technologies, such as social media

- The identification of safe passage routes and refuge centers for evacuated populations

- The identification of special-needs populations and how to effectively and efficiently move them

Protective actions—protect-in-place. This decision is based on the previous risk analysis determined by response agencies and the LEPC. This principle is grounded on the assumption that keeping the affected population in place as opposed to evacuation will provide a safer environment for them. During a protect-in-place action, people are directed to stay in a building, close all doors and windows, shut off HVAC systems, and not exit the structure until told to do so. This type of protective action may not be the best option if the vapors are flammable, it will take a long time for the gas to clear the area, or if the building(s) cannot be sealed tightly.

Protect-in-place is typically used during the following scenarios:

- The material is spreading too quickly, or it is too dangerous to risk exposure.

- There is a moderate- to low-hazard hazmat release.

- There are leaks that can be controlled quickly.

- Vapor clouds are expected to disperse rapidly.

- Hazardous materials have been completely released from their containers.

Some factors that can assist in determining this method of protection include the following:

- What is the potential threat, such as an airborne chemical or biological hazard?

- Will evacuation or protect-in-place provide the lesser danger to the public?

- Are there sufficient resources available to conduct a large-scale evacuation? If not, is protect-in-place the better option?

- Has the affected population been trained on the proper conduct when they are required to be in a protect-in-place situation?

Combination of evacuation and protect-in-place. Some instances may require a combination of evacuation and protect-in-place operations. This could be because some segments of the population would be safer being evacuated, whereas others are better off remaining indoors. This method is also used when on-scene resources are not sufficient to support a large-scale evacuation.

Methods to isolate area and deny entry

After determining the size of the endangered area, the first responder should take the following actions to secure the affected area and the indicated isolation distance:

- Position a responder in each area of approach.

- Direct traffic away from the scene.

- Use physical barriers such as barricades, vehicles, scene tape, and/or ropes; close doors and gates (fig. 3–77).

- Transmit warnings over vehicle or facility PA systems.

- Ask the media to make announcements to notify people to avoid the area.

Fig. 3–77. Use physical barriers to prevent access to the area.

Responders, especially those first on scene, must determine what methods of securing the scene will work best with the resources they have on hand. If additional resources are needed to properly secure the area, they should be requested immediately. Law enforcement can play a key role assisting with this process.

Establishing control zones

Control zones provide necessary scene control at hazardous materials incidents by regulating the movement of responders, thus reducing the potential for contamination.

Zones are set up to divide the levels of hazard at an incident and should be clearly marked (fig. 3–78). The size of each zone is based upon the degree of hazard that the material presents and requires the identification of the involved products. The zones may be adjusted based on changes in the incident. For example, wind should be considered when making the zone boundary determination.

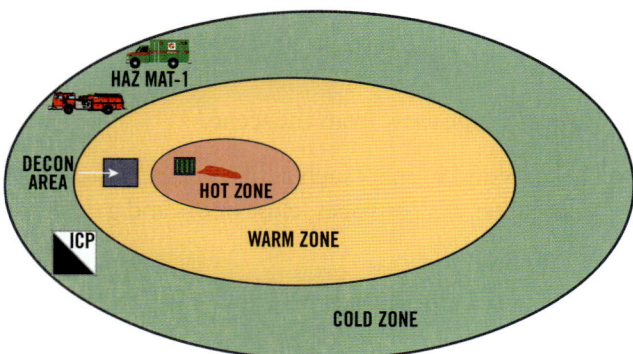

Fig. 3—78. Control zones are used to regulate the movement of emergency responders.

Similar to the methods used to determine the initial isolation areas and protective action distances, the initial determination of the scene control zone is based on information from the ERG, Chemical Transportation Emergency Center/Canadian Transport Emergency Center (CHEMTREC/CANUTEC), and scene observations and assessments. Long-term determinations are made by monitoring and sampling the materials and evaluating the extent of contamination, plume and dispersion models, and changes in weather or other conditions.

Hot zone. The hot zone is the contaminated area immediately surrounding the release. It should be large enough to prevent persons outside the zone from being exposed to the material, typically the size of the isolation distance and possibly including the protective action

area. The hot zone should include all areas exposed to any gases, vapors, dusts, or runoff of the hazardous material.

Work in the hot zone is performed by hazmat technicians wearing chemical protective clothing (CPC) designed for the substance released (fig. 3–79). Only essential personnel who are required to control the incident should be allowed to enter this area.

Fig. 3—79. The hot zone is restricted to properly trained responders wearing appropriate PPE at the technician level.

Incident control within the hot zone includes:

- Stopping the release

- Containing the released materials

- Triaging, removing, and decontaminating victims

Warm zone. The warm zone is the area between the hot and cold zone and is sometimes referred to as the decontamination reduction zone. It is used to support responders in the hot zone. Appropriate PPE may be required for responders working within the warm zone. Decontamination operations are conducted in this zone (fig. 3–80).

Fig. 3–80. Decontamination operations take place in the warm zone. The decon team are made up of Operations level responders.

Cold zone. The cold zone encompasses the warm zone. Personnel are not required to wear any special PPE in this area. The cold zone contains the command post (fig. 3–81), staging area, and treatment/triage area. Awareness level responders work in the cold/warm zone performing support tasks.

Fig. 3–81. The command post is located in the cold zone.

RESPONSE LEVELS

Incident levels

The response plan for a hazardous materials incident is based on the severity of the incident; for example, a response to a car leaking gasoline is less than an overturned tanker truck leaking chlorine. Hazardous materials response levels are usually designated Level I, II, and III and are based on the level of risk an incident involves. The LERP and department standard operating procedures (SOPs) should dictate the resources necessary for each level as well as mandatory notification procedures to other agencies.

Level I: potential emergency condition. Level I incidents are the least serious and most easily handled of the three levels. They can be controlled by the organization having jurisdiction without additional assistance. These incidents do not require evacuation outside of the structure or facility, and there is no immediate threat to life, health, the environment, or property.

The following are examples of Level I incidents:

- Gas leaks on the consumer side of the meter (fig. 3–82)

- Small diesel fuel or gasoline leaks from an automobile

- Leaks involving consumer commodities such as thinners, paint, bleach, swimming pool chemicals, and so on

Fig. 3–82. Natural gas leaks on the consumer side of the meter are typically considered Level I incidents.

Operations Response to Hazardous Materials

HAZMAT

Chapter 3

Level II: limited emergency condition. Level II incidents are beyond the capabilities of the first responders and require the response of additional resources, such as a hazmat response team. These incidents may require a limited evacuation of neighboring residents or facilities. They are not confined to a facility but remain within a confined limited evacuation of the surrounding area. First responders at a Level II incident are required to use chemical PPE.

- Response actions include the following:
- Diking and confinement of the material
- Plugging and patching operations
- Sampling and testing of unknown substances
- Implementing decontamination operations
- The following are examples of Level II incidents:
- Leaks or spills requiring large-scale evacuations
- Incidents involving extremely hazardous substances
- Fire with a potential boiling liquid expanding vapor explosion (BLEVE) threat
- Major incident involving a flammable liquid spill or overflow (fig. 3–83)
- Spill or leak of unidentified or unknown substances
- Underground pipeline rupture

Fig. 3–83. Major incidents involving flammable liquid spill are typically considered Level II Incidents. (Courtesy of William Seward)

Level III: full emergency condition. Level III incidents are the most serious of the three levels and require large-scale evacuations. These incidents may require resources from state and federal agencies as well as private industry and may necessitate the effort of several agencies to mitigate the incident. It is not confined or has moved past its confined geographic area and is an actual or perceived threat to life, health, and environment.

Response actions include the following:

- Use of specialized spill and leak control techniques
- Sophisticated sampling and monitoring equipment required
- Government and industry specialists required
- Use of large-scale decontamination setups

The following are examples of Level III incidents:

- Train derailment involving a leaking chlorine car
- Incidents resulting in the activation of the federal response system
- Incidents exceeding the capabilities of the local hazmat response team(s)
- Incidents requiring the evacuation of people extending across jurisdictional borders

Progress evaluation and reports. HM 5.6.1 Once hazards are determined and tactical decisions are made an Incident Action Plan (IAP) is formed. Team leaders assign tasks to team members and these tasks must be executed according to the IAP. As tasks are completed progress evaluations are reported to team leaders and then branch officers and finally to the incident commander. This allows the command staff to stay aware of incident conditions. Items that should be reported are;

- Task completion
- Did the task completion result in the desired outcome?
- Current condition of the incident

Command staff will take progress reports and review the action plan. This will continue until the hazard is mitigated. At the end of the incident or work period all responders at the incident should participate in a debriefing. Debriefings should be conducted immediately following the work period and be held in a quiet area the is distraction free and cover the following:

- Objectives that were completed
- Identity of the product

- Signs and symptoms of exposure

- Actions to take for exposures

Recovery. All defensive actions should be planned with future recovery operations in mind. Recovery includes getting equipment and personnel decontaminated, response units restocked and back in service, incident cleaned up. The cleanup can take a significant amount of time and is conducted by private cleanup contractors. It is not uncommon for incident scenes to take days to months to get the area back to pre-incident condition.

INCIDENT COMMAND SYSTEM

An **incident command system (ICS)**, also known as an **incident management system (IMS)**, is a crucial component in the organization of emergency incidents, including hazmat responses. It allows for more efficient operations and a more effective span of control (fig. 3–84). It can be developed locally or adopted from a nationally recognized system, and should become part of the local emergency response plan (LERP). The ICS should specify the roles and responsibilities of each position during various emergency responses.

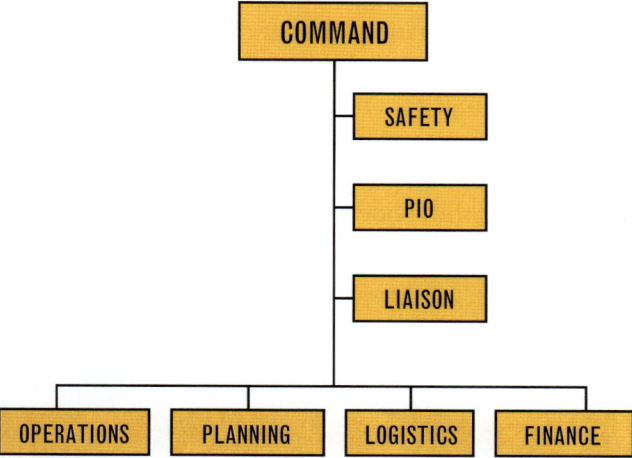

Fig. 3–84. The basic ICS chart

For the purpose of this section, only basic ICS components and those aspects related to hazardous materials responses are addressed. While NFPA Standard 1072 & 472 uses the IMS term and acronym in discussing the relevant requirements, we will use the ICS term throughout this section to maintain consistency with National Incident Management System (NIMS) terminology.

Elements of ICS

To be an effective scene management tool, the ICS must include the following components:

- **Common terminology**. Terms used to describe organizational elements, resources, and facilities must be consistent.

- **Modular organization**. The system should be designed from the top down with various branches/ sections added, depending on the size and complexity of the incident.

- **Integrated communication**. Communication systems must allow the various agencies involved to communicate with each other.

- **Unified command structure**. Each agency with jurisdictional responsibilities needs to be represented in the command structure.

- **Manageable span of control**. The recommended span of control is five to seven people per supervisor.

The actual size and complexity of the ICS is dependent on the size of the incident it is designed to handle.

Hazmat ICS positions, groups, and branches

Depending on the complexity of the incident, not all positions, groups, or branches in the incident command system may end up being filled. For hazardous materials incidents, however, there are some that must be filled. The following positions are all crucial to a hazmat response.

Incident commander. The incident commander is responsible for the overall operations at the scene of all incidents including a hazmat response. Incident commander duties are as follows:

- Designate a safety officer.

- Establish the site safety plan.

- Implement a site security and control plan.

- Identify the materials and conditions involved in the incident.

- Implement appropriate emergency operations.

- Ensure that proper protective equipment is used at all times.

- Set up a decontamination operation.

Safety officer. A safety officer is required at all hazardous materials incidents. The officer is responsible for the following:

- Monitor and identify potentially hazardous or unsafe situations.

- Stop any operation deemed unsafe immediately.

- Maintain communications contact with IC.

- Assist in incident planning.

- Identify potentially hazardous situations on the scene.

- Review incident action plan for safety concerns.

- Identify and correct unsafe situations immediately.

Hazardous materials branch or group. Most incidents will benefit from a standard application of ICS. Other types of incidents, such as hazmat, mass casualty, and so on, may require the use of more specialized functions within the operations section of the ICS organization. The specialized functions include the following:

- Hazardous material group supervisor

- Science officer

- Entry unit leader

- Decontamination unit leader

- Site access control unit leader

- Assistant safety officer, hazardous materials

Any resources directly involved with the hazardous material are supervised by one of the functional leaders or the hazardous materials group supervisor. When the incident is more complex, a hazardous materials branch director may be required for proper span of control and organizational management.

The three functional positions within the hazardous materials group (entry leader, decontamination leader, and site access control leader) require a high degree of control and close supervision.

Hazardous materials group supervisor. The hazardous material group supervisor is assigned to the operations section and reports to the operation section chief or the branch director (if activated). The hazardous materials group supervisor is responsible for:

- The implementation of the phases of the incident action plan (IAP) dealing with the hazardous materials group operations

- The assignment of resources within the hazardous materials group

- Reporting on the progress of control oprations and the status of resources within the hazardous materials group

- Direction of the overall operations of the hazardous materials group

Science officer. The Science officer is responsible to conduct research on the product. Determine physical and chemical properties and assist in selecting the following tactical decisions:

- Predicting the behavioral of the material

- PPE selection

- Mitigation tactics

- Decontamination

- Public protection options

Entry leader. The entry leader supervises all companies and personnel operating in the exclusion/hot zone along with directing all tactics and control of the positions and functions of all personnel in the exclusion/hot zone.

Decontamination leader. The decontamination leader supervises all operations in the contamination reduction/warm zone (with the exception of those handled by the safe refuge area manager). The decontamination leader also ensures that all rescued citizens, response personnel, and equipment have been decontaminated before leaving the incident (fig. 3–85).

Fig. 3—85. The decontamination leader supervises decon operations in the warm zone.

Site access control leader. The site access control leader controls the movement of all personnel and equipment between the control zones. This position is also responsible for isolating the exclusion/hot and contamination reduction/warm zones and ensuring proper routes and responsibility for the control, care, and movement of people in the safe refuge area.

Assistant safety officer–hazardous materials. The assistant safety officer–hazardous materials reports directly to the incident safety officer and coordinates with the hazardous materials group supervisor or hazardous materials branch director (if activated). This position is also responsible for the overall safety of assigned personnel within the hazardous materials group. In incidents involving multiple emergency activities, the assistant safety officer–hazardous materials does not act as the safety officer for the overall incident. An assistant safety officer–hazardous materials should be appointed when the scope of the incident requires it.

Locating the command post

The incident command (IC) post may be the first-responding vehicle until a formal location can be established. The IC post can be established in a conveniently located building, a specialty equipped vehicle, or a predetermined location in the facility preplan (fig. 3–86). The command post is always located in the cold zone. Always relay the location to dispatchers and responding units.

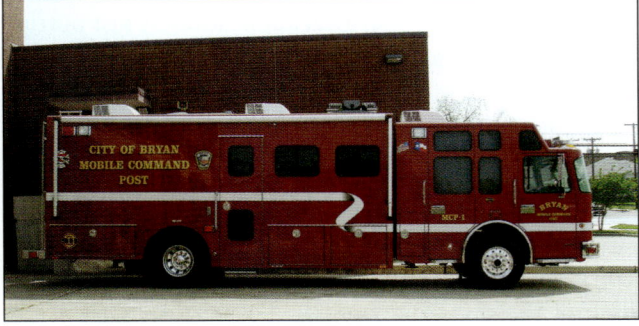

Fig. 3–86. A command post using a specialized vehicle

The following are factors to consider when locating a command post. It should:

- Be clearly marked
- Allow for controlled access to the site
- Be located uphill and upwind
- Be set up with the intention of not having to relocate it
- Allow for a view of the scene, if possible

FIXED FACILITIES AND FIXED FIRE PROTECTION

Fixed facilities have many types of fixed fire-protection, suppression, and detection systems. Remember that fixed systems are the first line of defense against fires, chemical releases, and other unwanted or unplanned events. Local codes dictate which type of facility is required to have which specific type of system; however, the responsibility for the installation and maintenance of the system lies solely with the owner/operator of the facility. In certain circumstances, owners will install a system at the request of their insurance company, or it may be a corporate policy to provide protection of certain processes to prevent major losses in the event an incident should occur, especially if that location is the only location that makes a particular product for the company. Businesses would rather spend the money on a fixed fire-protection system during start-up, rather than suffer a total loss and perhaps go out of business.

Automatic sprinklers

Automatic sprinkler systems include wet, dry-pipe, dry-deluge, preaction, and foam spray sprinklers.

Automatic wet sprinklers. With this type of sprinkler, water is right up to the sprinkler heads. They are used most commonly in heated spaces and are the quickest and most efficient type of sprinkler system. Locations: offices, labs, hospitals, home centers, schools, indoor garden centers, and lumberyards.

Automatic dry-pipe sprinklers. Compressed air holds back the dry-pipe valve until sprinkler heads start to open from the heat of the fire. The air rushes out of the heads, dropping the air pressure, which allows the dry valve to open and fill the system. Locations: outdoor lumberyards, home and garden centers, drum sheds, any other outdoor setting where freezing may be a factor.

Automatic dry-deluge sprinklers. This system contains all open sprinkler heads and can be triggered by an air pilot line, heat detectors, or a remote manual pull station. The pilot line had compressed air and closed sprinkler heads. When a fire occurs, the pilot heads open up form the heat, allowing a pressure drop. The air pressure holds a trigger pin on the valve in the closed position. When the pressure comes off, the pin releases and allows the deluge valve to open, delivering water to all heads, which are all open all the time. Locations: small tank farms, flammable liquid storage areas, high-

Operations Response to Hazardous Materials

HAZMAT

Chapter 3

voltage transformers, tank truck loading racks in terminals, outdoor chemical processing pads.

Automatic preaction sprinklers. This is also a dry system used indoors in water sensitive areas that takes two distinct actions to put water in the occupancy. The piping and heads are closed. There is *no* compressed air on the system. What actuates the valve are detectors in the room (heat or smoke).

Distinct action #1. When a fire occurs, the detection system in that area or room sends a signal to either a local fire alarm panel or a separate panel just hosting this system. The panel processes the signal and sends an electronic command to the preaction sprinkler valves telling it to open and fill the pipes. At this point, we now have a wet system waiting to go in service.

Distinct action #2. When the fire gets large enough to (fuse) activate or open the sprinkler heads, water will discharge on the fire. Where found: computer rooms, operating theaters in hospitals, water sensitive research equipment rooms or labs, medical testing buildings, such as nuclear magnetic resonance (NMR) and magnetic resonance imaging (MRI).

Automatic foam spray sprinklers. Similar in operation to the dry-deluge, dry-pipe, and preaction systems, except that this system has a tank of firefighting foam connected to it that delivers finished foam to the sprinkler heads. Where most sprinkler systems are designed to confine and control a fire, this system is designed to extinguish a fire. Where found: flammable liquid storage areas (drum pads, drum sheds), small tank farms with above-ground cylindrical tanks, and indoor or outdoor bulk processing using flammable or combustible liquids.

Foam systems. Foam systems consist of large-diameter pipes, foams tanks, and a water supply with pumps to deliver foam. Where found: large bulk storage tanks usually found in bulk oil terminals and refineries, oil loading or drilling platforms, petroleum tank loading docks and berths, and tank truck loading racks in terminals.

Fire alarms. Fire alarm systems consist of automatic and manual systems. Automatic systems require no human interaction, use detection devices, notify the occupants, and can notify emergency responders through electronic transmission. Detectors may include smoke, heat, flame, gas, and special industrial gases such as chlorine. Where found: most commercial or industrial occupancies and facilities. Manual systems require human intervention, are activated by pull stations, and can notify emergency responders through electronic transmission. Where

found: most if not all occupancies noted above with a few exceptions (e.g., grain silos).

Non-water-based systems

Halon and other clean agent suppression systems.
This is a colorless, odorless gas fire suppressant that interrupts the chemical chain reaction of burning. It is often used as a first line of defense against fire and can be installed with or without sprinkler protection. Halon is no longer in production because it is a chlorofluorocarbon (CFC) that disturbs the ozone layer and may be contributing to global warming; however, there are many halon systems still in service all over the country. Where found: computer rooms, medical testing buildings (e.g., NMR, MRI). With halon out of production, new environmentally friendly gas suppression agents have been developed. They are installed and work similarly to halon systems.

Dry chemical. There are two types of dry chemical. BC dry chemical is used for Class B (flammable and combustible liquids) and/or Class C fires (energized electrical). ABC dry chemical is used for Class A fires (wood, paper, ordinary combustibles) and for B and C fires as well. These systems can be activated by the melting of a fusible link on a spring-loaded line to a control head or by automatic detection. Once the control head activates, it allows the dry chemical to travel from its storage container through fixed piping to discharge nozzles. These systems can be used to total flooding (fill up the entire room) or for spot/local application by just putting the nozzles over a specific hazard within large room or area. Where found: flammable and combustible liquid processing and/or storage areas, under self-service gas station canopies, commercial cooking hoods. It should be noted that ABC dry chemical and BC dry chemical are not compatible with one another. They will tend to cancel each other out if used on the same fire at the same time.

Carbon dioxide (CO_2). This is a colorless, odorless gas that activates by detection or remote manual pull station by displacing oxygen to smother the fire. This system can dangerous to human life as it is normally designed to bring the level of oxygen in a room or area to below 15%. Where found: computer rooms under the raised floor, as well as flammable liquid dip tanks.

Table 3–5 shows which systems *may* be found in various fixed facilities, depending on local code requirements and or insurance company mandates.

Table 3—5. Fire Protection Systems in fixed facilities

Type of Fixed Facility	Fixed Fire Protection
Gas station	Dry chemical
Home center	Wet and dry sprinklers, fire alarms
Lumberyard	Wet and dry sprinklers, fire alarms
Bulk fuel depot	Foam, fire alarms, sprinklers
Refinery	Foam, fire alarms, sprinklers
Specialty chemicals	Various systems depending on the hazards
Hardware store	Sprinklers, fire alarms, dry chemical
Dry cleaning plants	Sprinklers, fire alarms
Garden centers	Wet and dry sprinklers, fire alarms
Photo processing labs	Wet sprinklers, fire alarms
Farm supply stores	Wet and dry sprinklers, fire alarms
High schools/colleges (labs)	Wet and dry sprinklers, fire alarms
Research labs	Wet and dry preaction sprinklers, fire alarms
Bulk pharmaceutical plants	Wet and dry sprinklers, fire alarms, foam, CO_2, foam spray, halon, preaction sprinklers, dry chemical
Water treatment plants	Dry deluge systems (vapor suppression)
Hospitals	Wet and dry sprinklers, fire alarms, halon, CO_2, preaction sprinklers

INTERNATIONAL FIRE CODE (IFC)

Chapter 27 of the International Fire Code discusses hazardous materials. Under section 2701.1, Scope, it reads, "Prevention, control and mitigation of dangerous conditions related to storage, dispensing, use and handling of hazardous materials shall be in accordance with this chapter." Table 3–6 lists the requirements set forth in the IFC for fixed facilities.

Table 3—6. International Fire Code (IFC) requirements

Section	Code Requirement
2701.3.3.1	Physical and health hazards must be made known to workers, neighbors, and the fire code official.
2701.3.3.2	Equipment must be designed, installed, and maintained for reliability purposes.
2701.3.3.3	Safeguards must be provided to prevent accidental release or unintentional reaction.
2701.3.3.4	Spill containment systems must be provided.
2701.3.3.5	There must be safeguards for ignition control.
2701.3.3.6	Hazardous materials must be protected from fires.
2701.3.3.7	Exposures must be protected from hazardous materials.
2701.2.2.8	Gas and vapor detection must be provided for quick response.
2701.3.3.9	There must be reliable power sources when power is required for emergency systems and mitigation.
2701.3.3.10	Means of ventilation shall be provided where required.
2701.3.3.11	Process hazard analysis must be performed to minimize an accidental discharge.
2701.3.3.12	A pre-start-up safety review must take place before starting a new or altered process.

Section	Code Requirement
2701.3.3.13	There must be a written plan for emergency action on the part of plant employees.
2701.3.3.14	There must be a written plan to manage change.
2701.3.3.15	There must be a written emergency plan for all employees.
2701.3.3.16	There must be a written plan for accident investigation and documentation.
2701.3.3.17	There must be an analysis completed for an off-site impact of a hazardous material release.
2701.3.3.18	Safety audits must be conducted periodically to maintain compliance with these sections.

The IFC also lists requirements for having a Hazardous Materials Management Plan (HMMP), a Hazardous Materials Inventory Statement (HMIS), and a plan for closing the plant, which must be submitted to the Fire Code Official 30 days prior to closing. The IFC lists quantity limits for storage of materials, as well as the design and construction of tanks, vessels, and containers that hold hazardous materials. Other sections include piping systems, equipment, building construction, identification systems, and fixed fire-protection systems.

Fixed facilities have a lot to do to be compliant with just the IFC. Add to that state and federal regulations, and it is easy to understand why industries have entire staffs of people dedicated to compliance alone.

PRE-EMERGENCY PLANNING FOR FIXED FACILITIES

Large industrial facilities in communities across the country are also known as target hazards. Not unlike preplanning a hospital or nursing home, pre-emergency planning a fixed facility is of paramount importance for the safety of not only the people who work there but also for any and all responding agencies. The priorities are always life safety, property protection, and protecting the environment. Industry adds protecting its assets and business continuity. By preplanning your fixed facilities, not only will responses be more effective but also every industry will something to offer the community, no matter how big or how small. However, you'll never know what that something is until you knock on the gate and open a dialogue. Responding lights and sirens to the local plant is great if you know the inherent hazards, and you have a formidable pre-fire plan to operate under. If you don't take your time, running in blind may injure or kill someone, and it could be you or your crew. Only through a good working relationship will lives be saved and the community will sleep easier at night.

How to start the preplanning process at fixed facilities

A meeting between the plant manager and the fire chief (or appointed fire official) and their staffs is the best place to start. From the plant may come the plant fire chief or safety professional, environmental professional, and security; from the municipality, the first due company officer, battalion or deputy chief whose district it is, and the fire-prevention official. Perhaps a cursory tour of the plant is in order at the first meeting, and some basic data to change hands, like a map of the facility showing entry points, high hazard storage areas, firefighting equipment (hydrant grid, pumps, sprinklers, standpipes, fire department connections, etc.), and a copy of the plant emergency response plan. If in fact they don't have a plan, this is another opportunity for the industrial and municipal personnel to get together to come up with something that's basic and functional. At the conclusion of this first meeting, each group should appoint a liaison who will then keep in touch and work to perpetuate the relationship. In most cases, it's the staff of the fire prevention bureau and the plant fire chief or safety professional. This may seem like a long road, but it's not impossible.

Building relationships

The fire prevention bureau in most jurisdictions around the country has the "right to inspect" under a local law of some type. The fire inspector can help bring the facility into compliance while picking up good information to bring back to the responding suppression companies. After the fire prevention bureau completes it inspection, it may be prudent that the plant manager invite the local and/or first-due companies to the plant for a tour. Once again, the firefighters will get an opportunity to see firsthand what the process or storage areas look like and can also examine the in-plant fire-protection systems. One tour won't do it. Each working crew must get into the plant to see what's there. Showing a crew of four from a fire station will not fit the bill. The plant must make the time commitment to get as many fire department personnel in and around as possible. This is especially important when there is no on-site brigade, fire department, or emergency response team. Tours should be convenient for both parties and set up for mutually convenient times. These types of activities can be expanded to other groups in the area (e.g. EMS, local hazardous materials teams, etc.)

Planning and sharing

Larger industries have a mid-manager-level crisis or emergency management team. This team is responsible for responding during a site-wide emergency (depending on the severity) and handling things that fall within the Incident Command System (ICS), like planning, logistics, finance, and the like. The team's primary role is to support the operations sector; disseminate information to the community, the press, and the corporate staff; and to plan for the recovery phase of the incident. Having the fire official from the community work with the plant's team will also enhance the relationship. This is another good opportunity for plant management and the fire department (and perhaps the local police as well) to work together and see what the site's plan calls for during specific types of emergencies (fire, explosion, chemical release, etc.). In many cases, the local fire department is asked to observe and participate in on-site drills and emergency response exercises. They are then asked to critique the plant's emergency plan for effectiveness and functionality. It's much too late to start planning for a disaster when the alarm is transmitted to the fire department or when the site security office gets the call on the site's emergency line. Both parties must make the time and take the effort to work together toward the common goal—which is the successful and safe conclusion of an emergency operation.

Joint training

Most if not all larger industries particularly chemical, petrochemical, refining, and similar type sites have some sort of on-site emergency response team. These teams vary, and some are incipient fire brigades, spill or hazmat teams, evacuation fire marshals, or full structural fire departments—some full-time paid and some volunteer. Many industries also have confined space, trench, and/or high-angle rescue teams and EMS capabilities as well. Whatever the type of team, joint training can only pay large dividends when an incident ensues and lends

itself to pre-emergency planning. Most often there are options where the two groups can practice on mockups in areas that are temporarily out of service or have been mothballed. Any joint training will certainly help to build not only the plant–municipal relationship but personal relationships as well. When an incident occurs, it always seems to work better when the two entities have trained together, and thus there is mutual respect among the responding personnel. Everyone will have a better idea of what one another's capabilities are in terms of equipment and training.

Staying with the process

Now that the relationship has been built, a plan must be developed to keep the momentum. Regular meetings at all levels should be scheduled, and they should include the plant manager, the chief, the brigade, the first-due company, the plant chief or safety professional, and the fire-prevention bureau staff. Joint training continues. Drills are held and critiqued. The plant emergency plan gets updated, and the prefire plan reworked if necessary. What about process changes, new construction, demolition, renovations, and changes in numbers of personnel? Is the plant map current, and are copies available to the response agencies? Is the plant policing itself for fire violations and good housekeeping, and thus building trust with the local fire official? Has the plant manager made the facility the center of the community, and is the company a good neighbor? These are all good questions to ask, and if the answers are negative in any way, it's time to take action.

Don't wait to start planning when things start to happen. It's too late. Make a concerted effort to work together and be prepared for whatever may come your way. A good pre-emergency plan and regular multi-agency drills will pay off when you get that once-in-a-career incident.

On-site surveys

The following is a list of things to consider when moving through a fixed facility while preplanning. Note that pre-emergency plans can get large and cumbersome. While all of this information is important and may be needed, consider a comprehensive plan for the chiefs and incident commanders, then extract enough information for first-arriving companies by creating a quick action plan (QAP). A QAP can be a one-page sheet with critical information on arrival.

- **Occupancy.** What is going on in that building? What is the fire load?

- **Life hazard.** What is the occupant load? How many people are working here and where are they?

- **Water supply.** Hydrants (hose thread), distribution mains (sizes), fire pumps, storage tanks

- **Building construction.** How were the buildings built? What types of construction are they? How many stories? What's the footprint?

- **Bulk storage/hazardous materials.** Locations, types, classes, amount, material safety data sheets (MSDSs).

- **Fixed protection.** Sprinklers, standpipes (hose thread), foam, CO_2, dry chemical, wet chemical, clean-agent systems, fire alarms, detection (smoke, heat, flame, special), fire alarm panels, explosion suppression systems, fire department connections.

- **Fire flow calculations.** How much water will we need for the hazards present?

- **Building systems.** Elevators (electric/hydraulic, firefighter service), HVAC (heating, ventilation, and air-conditioning), electrical, natural gas, fuel oil systems, domestic water, emergency generators, transformers, high-pressure steam, smoke control system.

- **Egress and ingress.** Inside: stairwells, doors, ramps, loading docks, fire escapes, fire towers. Outside: obstructions, overhead wires, bridges, overpasses, narrow access roads.

- **Exposures.** What's around the buildings that could aid in the spread of fire?

- **Passive fire protection.** Fire walls, fire doors, fire partitions, smoke barriers, fire barriers.

QUESTIONS

1. List three types of fixed facilities that may house hazardous materials.

2. Where are NFPA 704 placards usually found?

3. The Hazardous Materials Information System (HMIS) marking used by industry is similar to NFPA 704 except for an additional section that was added. What does that section identify?

4. What is the four-digit number placed inside of a DOT placard known as?

5. How many hazard classes are there in the DOT placarding system used for tank trucks, rail cars, and other vehicles transporting hazardous materials?

6. What are the main purposes or goals of the Global Harmonization System (GHS)?

7. What do the green section/pages of the DOT Emergency Response Guide tell first responders?

8. Shipping papers used for rail operations are known as _____ and are kept in either the _____ or the _____.

9. What are the best and most immediate two actions for an awareness-level responder to take upon arrival at a scene?

10. What do the blue section/pages of the DOT Emergency Response Guide tell first responders?

Hazardous Materials Operations: Mission Specific

OBJECTIVES

Upon completion of this chapter, you should be able to do the following:

- Explain the types of respiratory equipment used in hazardous materials emergencies.

- Explain the types of thermal protective clothing.

- Explain the types of chemical protective clothing.

- Understand the types of thermal and psychological stress encountered while wearing personal protective equipment.

- Describe the procedures for using personal protective equipment.

- Explain the inspection, storage, maintenance, and testing programs for personal protective equipment.

- Explain the types and methods of decontamination used during hazardous materials emergencies.

- Identify the procedure for conducting gross decontamination.

- Identify the procedure for conducting emergency decontamination.

- Identify the procedure for conducting, reporting and documenting technical decontamination.

- Identify the procedure for conducting mass reporting and documenting decontamination.

- Describe decontamination set up.

- Explain crowd management for mass decontamination.

- Identify agencies with evidence collection authority.

- Explain the implementation of scene security procedures.

- Identify agencies with evidence collection authority.

- Identify agencies with investigative authority.

- Describe the processes for the collection, protection and preservation of evidence.

- Describe the process for public safety sampling.

- Explain the unique hazards presented at crime scenes.

- Describe the documentation requirements for evidence collection.

- Explain concentration, dose and exposure levels.

- Explain the types of monitoring equipment.

- Describe the procedure for the selection, use, limitations and maintenance of monitoring and sampling equipment.

- Explain the types of sampling equipment.

- Identify hazards present at various illicit labs.

- Explain the operations at illicit labs.

- Explain the methods to remediate illicit labs.

- Describe the common indicators of illicit labs.

PERSONAL PROTECTIVE EQUIPMENT (PPE)

Types of personal protective equipment

HM 6.2.1 In addition to the competencies required to perform as a hazardous materials operations-level responder individual authorities having jurisdiction (AHJ) can choose to train responders to perform mission specific tasks. Operations level personnel trained to operate at the mission specific level for personal protective equipment must work under the guidance of a hazardous materials technician, an allied professional, an emergency response plan or standard operating procedure while selecting and using PPE.

Personal protective equipment is an important component in any hazmat responder's inventory. More important, however, is the knowledge of how to pick the appropriate level of protection for the incident. Failure to understand how to interpret hazard and response information and use it to choose the correct PPE can result in significant injury to response personnel. Responders should also be trained in the use of any specialized hazmat/WMD PPE issued by the authority having jurisdiction (AHJ). The two major categories addressed here are respiratory protection and personal protective clothing. Operations-level responders must be specifically trained to use chemical protective equipment (CPE) and in the operation of technical decontamination before they are permitted to wear CPE at any hazmat/WMD incident. Also, they must know the differences between the various levels of protection

and understand how to choose the appropriate level of PPE for a given material and how the available level of protection affects the options for various operations. If there is a lack of appropriately trained personnel and/or equipment, a request for personnel with the proper level of training and equipment should be transmitted immediately.

Respiratory protection. Respiratory protection equipment protects the first responder against the inhalation of toxic substances. The various types of protective breathing equipment are open-circuit self-contained breathing apparatus (SCBA), closed-circuit SCBA, supplied-air respirator (SAR), air-purifying respirator (APR), powered air-purifying respirator (PAPR). Each type of respiratory protection has its advantages and disadvantages. Because the respiratory risks for first responders at hazmat/WMD incidents vary greatly, selection of the appropriate respiratory protection is essential. If there is any uncertainty, responders should always default to using and SCBA or SAR. It is readily available to fire service personnel and provides the highest level of respiratory protection. OSHA mandates that SCBAs be utilized anytime oxygen levels drop below 19.5% (oxygen deficient) or above 23.5% (oxygen enriched). However, due to fact that many chemicals can cause injury or illness at minute levels responders should consider using respiratory protection even with the slightest changes in oxygen concentration. All SCBA utilized for hazmat/WMD incidents should meet NFPA 1981 and NIOSH certification for CBRN.

Open-circuit SCBA. The open-circuit SCBA is the most commonly used breathing apparatus in the fire service (fig. 4–1). Only positive-pressure SCBA can be used for hazmat/WMD incidents. Positive-pressure units provide the wearer with a greater level of protection by maintaining a slightly-higher-than-atmospheric pressure in the mask. This substantially reduces the potential for contaminants to enter the facepiece. In an open-circuit system, air for breathing is compressed and stored in a cylinder carried on the wearer's back. Exhaled air is vented to the outside atmosphere through the regulator-mask assembly. Components of an SCBA include, high pressure air cylinder, backpack frame, harness, regulator, and facepiece. Also, SCBA's are equipped with low air alarms and newer versions have a heads up display (HUD) so responders are better equipped to monitor breathing air levels.

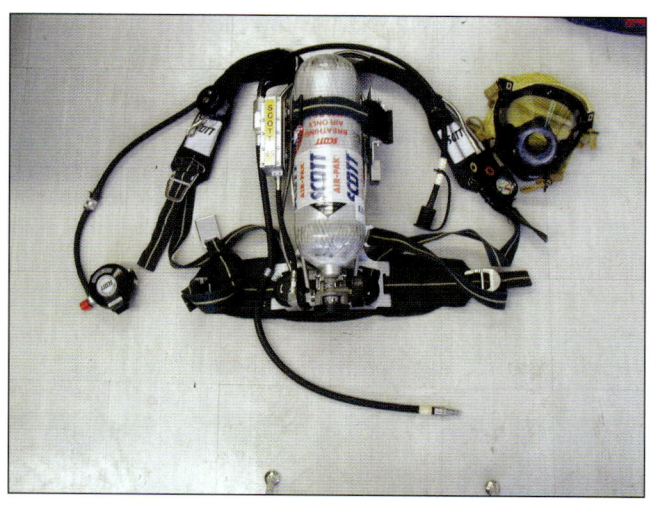

Fig. 4–1. Open-circuit SCBA is the most common type used in the fire service.

Air supply duration is dependent on the wearer's fitness, the degree of physical exertion, ambient temperatures, stress levels and the design of the unit. Personnel operating in the hot zone must remain cognizant of their air supply. Travel time from the hot zone to the decontamination line and estimated time to perform decontamination should be factored in when determining exiting air levels. Responders should never wait for the low air alarm before exiting the hot zone.

The advantages of SCBAs are that it allows for a greater range of mobility and provides the responder with the highest level of protection. The disadvantages include limited usage time, weight, and difficulty to use in confined spaces. Open-circuit SCBAs are available in 30, 45, 60 and 75-minute rated cylinders. Due to long distance to the hot zone and decontamination time responders at hazmat/WMD incidents use at least 60 minute rated cylinders.

Closed-circuit SCBA. A closed-circuit SCBA consists of a cylinder of oxygen, a filtering system, a regulator, and valves. The system works by cleaning and filtering exhaled air and then adding oxygen. Usage time for a closed-circuit SCBA ranges from 30 minutes to 4 hours.

The advantages of a closed-circuit SCBA include the duration of air supply that exceeds the open-circuit model. Also, it typically weighs less than open-circuit SCBA. A disadvantage is that it is more complicated to use than open-circuit SCBA and requires special training.

Supplied-air respirator (SAR). In a SAR, the airline is attached to a regulator assembly containing one or more cylinders (fig. 4–2). While SARs are not commonly used by the fire service for hazmat applications there are

certain situations where they are preferred. For instance, depending on the chemicals involved, decontamination lines can operate much more efficiently with SARs. If responders are working in an IDLH atmosphere the SAR assembly must come equipped with an emergency breathing support system (EBSS), also called an escape pack, which is a small breathing air cylinder rated for 5–15 minutes. This allows the responder to escape a hazardous atmosphere in the event of an emergency. A SAR has a multi-cylinder cart or compressor system that connects to the airline manifold. This manifold allows multiple responders to operate off one cart. The airlines are connected to a standard open-circuit face-piece, a regulator, and if necessary, an escape pack. No more than 300 ft. length of air line is permitted for use with each user.

Advantages of SARs include increased working time, the ability of a first responder to work without a cumbersome backpack assembly, and that it is lighter than SCBA. Disadvantages include limited travel distance away from the regulated air source, the potential for the wearer to become entangled in the air line, and the hose possibly coming in contact and reacting with the hazardous material.

Fig. 4–2. The supplied air respirator is not commonly used in the fire service for hazmat applications.

Air-purifying respirator (APR) and powered air-purifying respirator (PAPR). Both APRs and PAPRs work by filtering ambient air (fig. 4–3). They come in full face or half face models and cartridges capable of filtering particulates and gas/vapor contaminants. The difference between an APR and a PAPR is when using an APR, the wearer draws air through the cartridge by breathing while a PAPR has a blower that pushes air through the cartridge. As the user becomes fatigued drawing through the cartridge becomes more difficult and reduces user work time. There are many restrictions tied to the use of APR's and PAPR's that make it difficult to use them in emergency response. Restrictions for APR/PAPR use are as follows:

- Each cartridge is manufactured for a specific chemical or group of chemicals

- Cannot be used in an IDLH atmosphere

- Cartridges have a shelf life and must be discarded once exceeded

- Cannot be used in oxygen deficient (19.5%) or oxygen enriched (23.5%) atmospheres

Fig. 4–3. Air purifying respirators work by filtering the ambient air and are not typically used in emergency situations.

The advantages of APRs and PAPRs are that they are lightweight, easy to use, and allow for greater mobility. However, they will not

Types of personal protective clothing (PPC). As

with respiratory protection, PPC is a key component in a responder's protective ensemble. Operations-level responders need to be aware of the different levels of PPC available to them as well as the purpose, advantages, and limitations of each of the options. Always keep in mind that no single type of PPC will protect a responder in all

situations. PPC is necessary to protect the skin and the body from contact hazards such as chemical burns and the potential for absorption of toxic materials through the skin.

Work uniforms. Work uniforms are worn on an everyday basis and offer the least amount of protection in a hazardous materials emergency (fig. 4–4). Work uniforms may prevent day-to-day dirt and grime from getting on skin, but they offer no mechanical, thermal or chemical protection.

Fig. 4–4. Work uniforms offer the least amount of protection at a hazmat incident.

Structural firefighting protective clothing (SFPC). SFPC is used every day for firefighter responses (fig. 4–5). When worn in combination with SCBA, it primarily offers protection from thermal and mechanical harm. Structural firefighting protective clothing is not recognized as chemical protective clothing but can provide limited protection against some chemical exposure for short duration tasks. These tasks should be limited to life saving activities or necessary public protection activities. However, responders must try to eliminate any openings in the SFPC and skin exposure. Also, they should make every effort to stay out of the product or limit that exposure. Any responders that enter the hot zone in SFPC should undergo decontamination even if they don't believe they contacted the material. Very little chemical exposure testing on SFPC has been accomplished. Chemicals can permeate, penetrate or degrade SFPC exposed to chemicals. Therefore, complete decontamination should be accomplished and disposal should be considered if any SFPC that has experienced chemical permeation. Also, SFPC may be used while performing defensive operations or support functions in areas where the levels of toxicity are very low. SFPC usually consists of SCBA, personal alert safety system (PASS) device, helmet, bunker coat, bunker pants, boots, gloves, and protective hood.

Fig. 4–5. Structural PPE provides thermal protection to the responder but is not considered chemical PPE

High-temperature protective clothing. High-temperature protective equipment is a level above structural firefighting protective clothing (fig. 4–6). This type of protective clothing shields the wearer during short-term exposures to high temperatures. High-temperature protective clothing consists of approach suits, proximity suits, or fire entry suits. An approach suit allows firefighting operations to take place with high levels of radiated heat. A proximity suit allows for close approach to fires for rescue and suppression operations. Fire entry suits allow the wearer to work in total flame environments for short periods of time. High-temperature protective suits do not provide chemical protection and users are susceptible to heat stress because high-temperature clothing holds the bodies heat in. They are bulky and cumbersome to use, they also require specialized training and have limited applications.

Fig. 4–6. High-temperature protective clothing shields the wearer during short-term exposures to high temperatures. (Courtesy United States Air Force)

Chemical-protective clothing (CPC) and equipment. CPC is designed to prevent chemicals in solid, liquid, or gas form from coming in contact with the wearer. Each ensemble is designed for a specific level of protection, and the material that it is constructed of offers varying levels of resistance. The two main types of CPC are vapor protective and splash protective suits that are worn with either an SCBA, SAR or APR/PAPR. CPC provides no protection against thermal or mechanical hazards. Wearers must receive adequate training in the use of CPC. In many cases users experience an added level of claustrophobia when wearing CPC. This can possibly be controlled through extensive training with the garments. The golden rule of chemical protective clothing is very simple: there is no one chemical protective ensemble or material that will protect you from everything.

The U.S. Environmental Protection Agency (EPA) established the Level A, B, C, and D standards for chemical protective equipment. These levels are also recognized by NIOSH, OSHA, and the U.S. Coast Guard. The NFPA recognizes Levels A, B, and C.

The four levels of chemical protective clothing are summarized in table 4–1.

Table 4–1. Chemical protective clothing levels

Level A	Encapsulating suit that covers both the wearer and SCBA
Level B	Consists of a garment designed for splash protection and an SCBA
Level C	Level B splash protection garment with a respirator instead of SCBA
Level D	Ordinary work clothes or uniform and no respiratory protection

Level A ensemble. The Level A suit is fully encapsulating and covers both the wearer and the SCBA (fig. 4–7). It affords the highest level of skin and respiratory protection for the wearer. The Level A suits are designed to be used primarily when materials present a severe threat of injury from contact with vapors or skin absorption. This ensemble also meets the requirements that are described in the NFPA 1991 standard.

Hazardous Materials Operations: Mission Specific

HAZMAT

Chapter 4

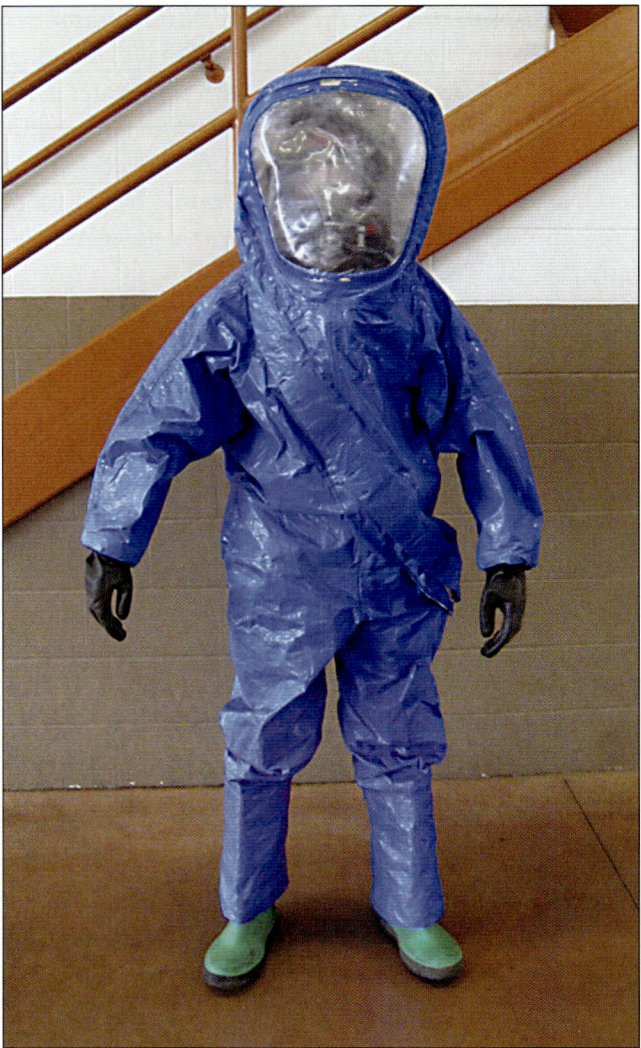

Fig. 4—7. Level A ensemble

The following are the typically recommended components of Level A protection:

- Fully encapsulating vapor-protective chemical-resistant suit

- Inner and outer chemical-resistant gloves

- Hard hat

- Communications system

- SCBA or SAR

- Chemical-resistant safety boots (including steel shank and toe)

- Overboots

- Level A protection provides the highest level of skin and respiratory protection against chemical vapor and splash. Disadvantages include no thermal protection, a potential for heat stress and impaired movement, communication, and visibility.

Level B ensemble. A Level B ensemble consists of a garment designed for splash protection and SCBA (fig. 4–8). This type of protection is used when it is unnecessary to provide protection from vapors. Potential areas of entry such as wrists, ankles, facepiece, and hood are sealed to prevent exposure from splashes. Responders will often use Level B protection during product confinement operations and while staffing decontamination areas. There are three types of Level B suits: two-piece, coverall, and encapsulating. The encapsulating suit is the type most commonly used by hazmat teams because it provides an increased level splash protection and reduces the risk of SCBA contamination. It provides protection against splash hazards but not against chemical vapors or gases. This ensemble also meets the requirements that are described in the NFPA 1992 standard and part of the NFPA 1994.

Fig. 4—8. Level B ensemble

The following are the typically recommended components of Level B protection:

- Chemical-resistant splash-protective suit

- Hard hat

- Inner and outer chemical-resistant gloves

- SCBA or SAR

- Communications system

- Chemical-resistant safety boots (including steel shank and toe)

- Outer boots

- Hearing protection, if needed

Disadvantages of Level B suits include no thermal protection, impaired movement, communication, and visibility. Additionally, because SCBA may be located on the outside of the suit, chemical compatibility issues must be considered.

Level C ensemble. In a Level C suit, the same type of garment is used as in Level B protection. This level allows for respiratory protection other than SCBA (fig. 4–9). The respiratory protection used in this level consists of APR or PAPR. The use of an air purifying respirator instead of SCBA changes the level from B to C. Air-purifying respirators cannot be used with encapsulating splash protection suits. A coverall suit is the style used in this application. This type of protection is not used unless the specific material is known and can be measured. Air purifying respirators have limited applications and should not be used in emergency response situations. This ensemble also meets the requirements that are described in the NFPA 1992 standard and part of the NFPA 1994.

The following are the components of Level C protection:

- Chemical-resistant splash-protective suit

- Hard hat

- Inner and outer chemical-resistant gloves

- APR or PAPR

- Communications system

- Chemical-resistant safety boots (including steel shank and toe)

- Outer boots

- Hearing protection, if needed

The advantage of the Level C ensemble is the increased mobility because SCBA is not required; the disadvantage is that it cannot be used unless the specific material is known, can be measured and continuous air monitoring is required.

Fig. 4–9. Level C ensemble

Level D ensemble. Level D clothing, consisting of ordinary work clothes or uniforms, is not adequate for first responders at a hazmat incident (fig. 4–10). It provides protection against workplace hazards, but not chemicals. This level does not require respiratory protection. As a result, it should be used when respiratory protection is unnecessary and there is no splash hazard.

The following are the components of Level D protection:

- Work clothes or uniform
- Hard hat or helmet
- Chemical or work gloves
- Safety glasses or other eye protection
- Chemical-resistant safety boots (including steel shank and toe)

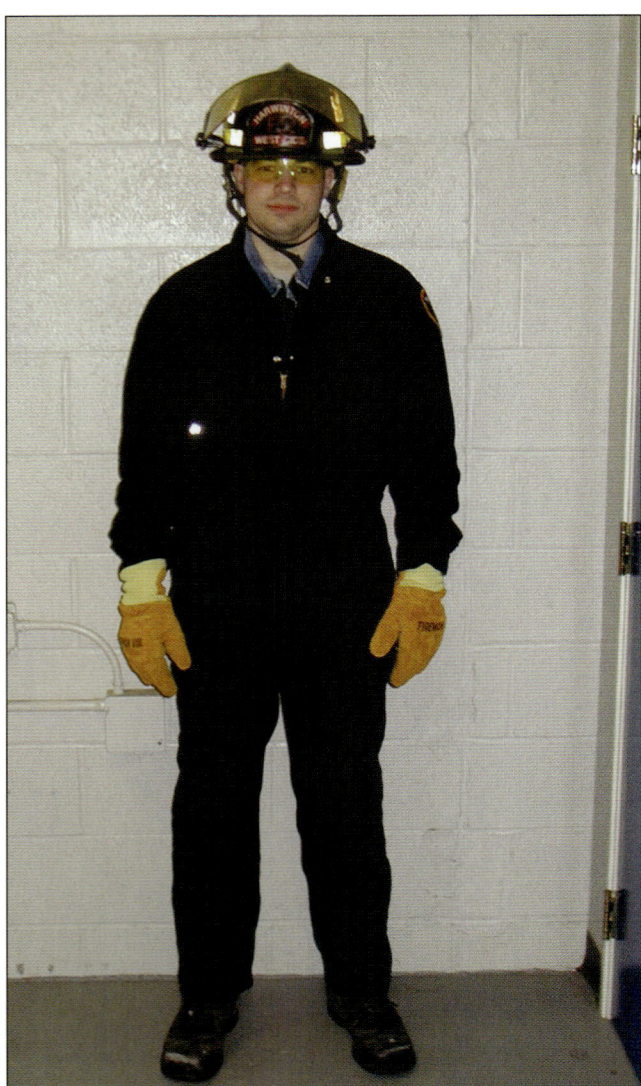

Fig. 4–10. Level D clothing

Personal cooling technologies

Working in personal protective equipment, especially chemical-protective clothing, can put a significant strain on the body, including the potential to overheat. Several different types of heat exchange units can be used to cool responders and reduce this risk. These include air cooled, ice cooled, water cooled, and phase-change cooling technology.

Air cooled. This is accomplished by supplying ambient or cooled air to the inside of suit through a hoseline. Advantages include cooling the surface of the skin. The disadvantages are a reduced mobility from the hose, limited range, and added weight

Ice cooled. This consists of ice packs worn in a vest on the inside of the suit. The advantage of this system is that it cools the skin. Disadvantages include added weight and unknown efficiency.

Water cooled. This uses a vest filled with chilled water worn on the inside of the suit. Some types utilize hoselines or tethers. Like air and ice cooled, the advantage to this method is that it provides cooling to the skin. The disadvantages are reduced mobility, added weight, and a short work cycle.

Phase-change cooling technology. This system uses a vest worn inside the suit and uses cool packs as a cooling medium. The cool pack material acts as a heat sink to draw heat away from the body. The material is designed to maintain a 59°F (15°C) temperature throughout the work cycle. Advantages include a lack of a shock to the system, such as when ice packs are used, a longer work cycle, and less time to regenerate than ice or gel packs.

Written PPE management programs

A written PPE management program is required for any agency that uses CPE in the response to hazmat/WMD incidents. Written programs include the following sections:

- Care and maintenance of PPE
- PPE Selection
- Training
- Medical monitoring

As with any equipment used in an emergency response, the regular care and maintenance of PPE is very important. Simply put, a responder's life depends upon this equipment functioning properly when it is needed. Manufacturer recommendations and department SOPs should dictate the procedures for cleaning, disinfecting, inspection, maintenance, testing, and storage. Personnel responsible for these procedures need to be well versed in the appropriate procedures.

Maintenance, testing, inspection, and storage

As with cleaning and disinfecting, the procedures for maintenance, testing, inspection, and storage are largely manufacturer-dependent. All equipment should be maintained, tested, inspected, and stored according to department SOPs and manufacturer's recommendations (fig. 4–11). Maintenance and testing of chemical protective equipment typically involves training and equipment that exceeds the operations level. When inspecting CPC, be sure to look for any abrasions, discolored areas, cracks, or other physical damage that would indicate damage. CPC should be stored in a manner that does not create any stress points in the material and in a container or location that protects against physical damage. Structural firefighting clothing and SCBA should be inspected and stored consistent with department SOPs. All required testing and inspections must be recorded and kept on file until the suit is discarded.

Fig. 4–11. PPE should be maintained and inspected in accordance with department SOPs and manufacturer recommendations.

Selecting appropriate PPE

HM 6.2.1 When choosing the appropriate PPE, the first responder should determine the identity of the material involved and the exposure type (skin, respiratory, or ingestion). Local protocols and department SOPs/SOGs will define the level of training, capabilities, and PPE available to operations-level personnel. The type of PPE available is an important factor in determining which type of action to use. Response personnel should always use their department's SOPs or SOGs to dictate the selection of available PPE based on the task at hand. The proximity of the responder performing his or her

duties to the release is also a factor. If the duties being performed are some distance from the release, such as building retention basins or building dams downstream to prevent the further spread of the material, then they need only to wear the appropriate safety equipment for that action (fig. 4–12). In other instances, such as responders operating closer to the spill to spread absorbent materials, additional PPE may be required.

There are several considerations that go into the selection of the correct CPC. The type of suit selected needs to provide chemical resistance to the materials the responder could potentially come in contact with.

Fig. 4–12. If the responder is operating in an area distant from the spill, chemical protective clothing is not likely to be necessary.

Chemical resistance is the ability of the garment to resist damage from direct contact with a chemical. As with chemical compatibility, several factors affect the resistance of the materials, such as time, temperature, stress, and reagents. CPC materials are specifically designed to resist the movement of chemicals into and through the material by one of three methods: permeation, penetration, or degradation.

Permeation is the process by which a hazardous chemical moves through a given material at the molecular level. Permeation occurs when the chemical absorbs into the material, moves at the molecular level through the material, and then de-absorbs to the interior. It then moves to the clothing or skin of the responder wearing the garment. All materials used for chemical protective clothing are tested against numerous chemicals to determine the amount of time it takes for them to permeate through the material. This is known as breakthrough time and manufacturers will document them on permeation tables for each garment they sell.

Penetration is the flow of a hazardous chemical through openings in the material such as zippers, seams, porous materials, pinholes, or other imperfections in the material.

Degradation is the physical destruction or breakdown of the material after being exposed to a chemical; it may consist of visible signs such as charring, shrinking, swelling, glazing, color changes, or dissolving. It may be at the molecular or nonvisible level and can be observed only by testing for weight of material, fabric strength, and other properties.

The PPE options for specific hazards

Thermal. For situations involving a thermal hazard, responders should select structural firefighting clothing, high-temperature clothing such as a proximity suit, or use flash protection for chemical-protective clothing. The type chosen depends on the level of thermal hazard present, available personal protective clothing, and whether there are any splash or vapor hazards associated with the material.

Radiological. Standard firefighting PPE offers only limited protection from radiological sources. The best protection is to determine the type of radiation present and use the time, distance, shielding concept discussed earlier to limit exposure and dosage.

Asphyxiating materials. For both simple and chemical asphyxiates, respiratory protection is the primary means of protecting response personnel. This can be worn in conjunction with structural firefighting gear and SCBA unless the material also presents a hazard requiring chemical protective clothing.

Chemical. Chemicals can present a splash hazard and a potential vapor hazard. Depending on the properties of the materials, the responder should select a Level A or B ensemble.

Etiological/biological. Protection from microorganisms is based on the means by which they would enter the body, their physical state, and how dangerous they are. For example, anthrax in a powder form would present a different hazard from a substance in an aerosol form.

Mechanical. PPE used by hazmat responders provides minimal protection from mechanical hazards. The best method to protect against this hazard is to avoid situations where a mechanical injury may occur.

Once the type of hazard is determined and the chemical and physical properties of the substance has been analyzed, command staff (science officer, safety officer and IC) can make the appropriate selection. If the only hazard present is thermal then structural firefighting or high temperature protective clothing may be the best selection. If there is a vapor skin contact or absorption hazard is present then level A would be your best option but if there is only a splash hazard then level B or C would be appropriate. However, sometimes you find that substances have more than one hazard such as a highly toxic, skin absorbent, highly flammable hazard. In this case, due to its toxic skin absorbent property a level A is necessary but due to its flammability level A does not provide adequate protection. In these cases, responders must eliminate one of the hazards before proceeding. One way to do this is through vapor suppression.

Training

Users must be trained before donning and using PPE for hazmat/WMD incidents. Initial training should include inspection, donning and doffing procedures, working in every type, emergency and technical decontamination procedures. In addition to initial training personnel are required to have, at a minimum, annual refresher training on the use of hazmat/WMD PPE.

Medical monitoring

All hazmat/WMD responders are required to have preemployment physicals as well as annual physicals. Also, all personnel donning CPE during an incident are required to complete a pre-entry medical screening prior to donning the CPE. This screening should include at a minimum; blood pressure, pulse, temperature and weight. Immediately following decontamination personnel must complete a post-entry medical screening. This screening should include blood pressure, pulse, respirations, hydration level, mental status, temperature and weight. Agencies should have a written medical monitoring policy to establish baselines for the medical screenings.

Any personnel who has been exposed or begins to exhibit signs and symptoms of exposure should report this to the designated exposure control officer. All exposure information including who to report exposures to will be discussed at the pre-entry safety briefing. All exposures should be recorded on an exposure record. Exposure records must be kept on file by employers for thirty years after the termination of employment. For instance, if you are exposed today and retire in 15 years then your exposure must be kept on file for 30 years after your retirement or 45 years.

Donning, doffing, and use of CPE

Personnel must assemble all the appropriate respiratory and CPE and conduct operational checks and inspections. Operational checks and inspections must be accomplished on SCBAs, SARs, APRs, and PAPRs. During these checks personnel should check for proper seals, on SCBA's and SAR's check that air cylinders are full, regulators are functional, low air alarms are operational and that there are no air leaks. For APRs and PAPRs check for seal, that the appropriate cartridge has been selected and that it is not expired. As discussed earlier, all personnel must go through pre-entry medical screening prior to donning CPE. CPE should be inspected for tears, rips and discoloration also that valves, zippers and seams are intact. Radio equipment and sound checks must be accomplished. The at least emergency decontamination has been setup (preferably technical decontamination). All monitors, tools and all equipment being taken into the hot zone should be assembled and inspected prior to donning.

An official safety briefing should be conducted prior to making entry. The only time this is not accomplished is if an emergency rescue is being performed. The safety briefing should cover:

- Incident description (including entry and egress routes)

- Container and its condition

- Product (if known)

- Signs & Symptoms of exposure

- Exposure procedures

- Entry objectives (recon, leak/spill control, rescue etc.)

- Emergency procedures (CPE failure, hand signals etc.)

- Use of the buddy system and an emergency back-up team

Once the PPE has been assembled, safety briefing concluded and pre-entry medical screenings are complete personnel should proceed to an open area to begin donning. The area should be under a tent or awning to keep personnel out of the elements if possible. If personnel to assist donning are available they should be directed to the donning are. Thin Tyvek suits can be worn as an undergarment instead of uniforms or personal clothing. Personal items such as wallets and jewelry should be bagged and secured. This allows for personnel to have dry clothing to change back into after the operational period. The entry team leader should supervise the CPE donning of the entry and back-up rescue teams while the decon leader should supervise the CPE donning of the decon team. It is important for personnel to stay together in their state of dress. This avoids members being completely dressed and others only beginning. Once CPE is donned a ready report should be communicated to the hazardous materials officer. Personnel should wait until given the order to make entry to go on air and zip up their suit.

Personnel entering the hot zone must diligently on keep track of their air supply. Hazardous materials response should be part of an agencies air management program. Prior to making entry the following actions should be factored into the air usage:

- Distance to the hot zone and return

- Decontamination (wet/dry, how many wash/rinse stations etc.)

- Type of tasks being accomplished (recon is easier than climbing on a railcar)

Personnel entering the hot zone should understand that CPE is your last line of defense not your first. In other words, they should make every effort to stay out of the product or limit contamination as much as possible. An increase in contamination means an increase in the chance of personal exposure. Personnel should not kneel on the ground as this can cause a decrease in suit integrity. Personnel wearing personal protective clothing during hazardous materials emergencies must undergo decontamination. Decontamination procedures will vary based on the materials encountered and the type of decontamination deemed appropriate. Detailed information pertaining to decontamination can be found next (section 6.3: Mass Decontamination and 6.4: Technical Decontamination) or in NFPA 1072.

MASS AND TECHNICAL DECONTAMINATION

HM 5.5.1 Decontamination is a process of removing or neutralizing contaminants that have accumulated on personnel, equipment, and the environment (fig. 4–13). Decontamination protects responders from agents that may contaminate and permeate the protective clothing, SCBA, tools, apparatus, and other equipment used at the incident. It protects all on-scene responders by minimizing the transfer of harmful materials to clean areas and helps prevent the mixing of incompatible chemicals. It also protects the community by preventing uncontrolled transportation of contaminants from the incident. Decontamination operations can be hindered by the lack of a positive identification of the material, weather conditions, topography, and availability of equipment and trained personnel.

Fig. 4–13. Decontamination is the process of removing contaminants from people or equipment. (Courtesy U.S. Air Force)

Common terms related to decontamination

Response personnel should understand the following terms and definitions as they relate to contamination concerns and be able to differentiate between them.

Contamination. It is the transfer of a hazardous material out of the hot zone in quantities greater than those deemed acceptable and which pose a risk outside the hot zone.

Direct contamination. Becoming contaminated by direct contact with the hazardous materials present in the hot zone, tools used by hazmat technicians becoming contaminated, or vehicles driven through a contaminated area. It can also be the result of the movement of the dusts, particles, gases, vapors, fumes, mists, and runoff

of the hazardous material and smoke and products of combustion spreading outside the hot zone.

Additionally, responders become contaminated by the following actions:

- Walking through a spill
- Touching the material
- Coming in contact with a vapor cloud

Among the basic methods of controlling the spread of contamination is setting up hot, warm, and cold control zones (fig. 4–14). People or equipment should not leave the hot zone without first being decontaminated. Access to control zones should be restricted. All contaminated items and clothing should be properly decontaminated or bagged and disposed of. The first responder should control all runoff from decon operations. If the release is indoors, the first responder should shut down HVAC systems to prevent other portions of the building from becoming contaminated.

Fig. 4–14. Control zones are set up to prevent the spread of contamination. A rail car such as this contains thousand of pounds of liquid which can rapidly turn to gas spreading over a large area in a short amount of time.

When dealing with radioactive materials, in addition to the previously noted methods, the first responder should employ radiation detection equipment to monitor all personnel and equipment leaving the decontamination area.

Cross-contamination. Is the contamination of people, the environment, and equipment outside the hot zone. It results from coming into contact with contaminated personnel, equipment, or property. Done properly, emergency and other decontamination operations prevent secondary contamination.

An example of secondary contamination is a victim of a hazmat incident who is transported to the hospital without proper decontamination. This endangers the EMS personnel and the emergency room staff, possibly resulting in their contamination.

Exposure. When exposed, a person has been subjected to a hazardous material via any of the various routes of entry into the body. This is different from contamination, in which any quantity of a hazardous material physically remains on people or objects. Exposure to a material does not necessarily cause contamination. For example, you may be exposed to a material in your workplace that meets the threshold-limit value for the time-weighted average (TLV-TWA) for that product. Being contaminated does not necessarily mean that you are exposed. For example, a responder's PPE may become contaminated at an incident; however, the responder is not exposed to the material unless the PPE fails or he or she is not properly decontaminated.

An exposure's hazard level depends on the material's concentration and length of exposure. Spills of a less hazardous substance require greater quantities to reach a harmful exposure level. Spills of **extremely hazardous substances** may only need a small quantity to cause a harmful exposure level.

Exposure from radiation can be either internal or external. Internal exposure is the introduction of radioactive materials internally to body cells, tissues, and target organs. Internal exposure cannot occur unless the victim has been contaminated. External exposure occurs when all or a part of the body is exposed to radiation from an external source. During an external exposure, the radiation may be absorbed by the body or it may pass totally through.

It is possible to be exposed to radiation but not be contaminated:

- Exposures can occur when a person is subjected to the particles or waves being emitted by the radiation source, without actually touching it.

- Contamination can occur when a gaseous, liquid, or solid radioactive material is released into the environment and deposits on people externally.

Routes of exposure

Hazardous materials may enter the body through four different routes of exposure: **Inhalation** occurs when chemical fumes, mists, dusts or gases are breathed in through the nose or mouth. The inhaled chemical is then absorbed through the tissue and membranes in the nose, trachea, and lungs. Tissues in these areas are not very protective against chemical exposure, thus are at greater risk. **Absorption** occurs when a hazardous material enters the body through the skin or eyes. Skin tissue is more protective than lung tissue, but is not an impermeable barrier. Some materials may be absorbed more readily by the skin than others, and once the material is absorbed, it is carried throughout the body in the bloodstream. **Ingestion** occurs when chemicals enter the body through the mouth and swallowed. Hazardous material is commonly ingested when contaminated food or hands come in contact with the mouth. **Injection** occurs when contaminated sharp objects puncture the skin, introducing hazardous material to the bloodstream.

Types of decontamination

Decontamination is the removal of hazardous contaminants from personnel, equipment, property, and the environment. There are four types of decontamination: gross decontamination, emergency decontamination, technical decontamination, and mass decontamination. Incident condition determines what types will be used. Decontamination can be grouped into two large categories: wet or dry. Wet decontamination requires the use of some sort of liquid, primarily water to wash or flush contaminates off people or objects. Wet decontamination often requires a significant amount of equipment and calls for control and disposal of runoff. Dry decontamination usually involves the removal of contaminants by brushing, scraping, vacuuming, absorbing and using adhesives to lift contaminants off people and objects. In addition to the two categories there are two methods chemical and physical. Physical decontamination is the removal of contaminants through a physical means:

- **Absorption** is the process of using absorbent materials such as earth, saw dust or vermiculite to collect the hazardous materials. Responders simply spread the absorbents over the materials and allow them to be absorbed then they can be collected. Once the materials are collected they can be disposed of. It is important to ensure compatibility between the contaminants and the absorbents used. Advantages are absorbents easy to get and usually inexpensive. Disadvantages are the material retains its hazardous properties, expensive to dispose of and you must ensure chemical compatibility.

- **Adsorption** is the process where hazardous materials adhere to adsorbent materials such as activated charcoal. Unlike absorbents, hazardous materials stick to the surface of adsorbents. Once the material is collected it can be disposed of. It is important to ensure compatibility between the contaminants and the adsorbents used. Advantages are absorbents are more effective and usually have less off gassing than absorbents. Disadvantages are the material retains its hazardous properties, adsorbents are expensive and you must ensure chemical compatibility.

- **Brushing and scraping** Is the process of removing solids off people and objects by using brushes and scrapers. Advantage is the equipment is inexpensive and easy to come by. Disadvantage is this method always leaves residual contaminants and must be followed up with another decontamination method, material retains its hazardous properties and must be disposed of.

- **Isolation and disposal** is the process of taking contaminated items, isolating it by placing it in a container such as an overpack drum and disposing of the materials at approved hazardous waste management sites. Advantage is easy to accomplish. Disadvantages are material retains its hazardous properties and disposal is expensive.

- **Vacuuming** is the process of capturing contaminants by use of specially designed vacuums. Advantage is highly effective in capturing dusts. Disadvantages are requiring special and very expensive vacuums, also requires expensive disposal as well as it might leave residue requiring additional decontamination methods.

- **Dilution** is the process of removing contaminants off people and objects by rinsing with copious amounts of water. If the substance is water soluble it may help reduce concentration of the material. Advantage is water is readily available and inexpensive. Disadvantages are containment of runoff, the material usually retains its hazardous properties, disposal is expensive.

- **Washing** is the process of using soap and water to suspend liquid and solid substances by the action of emulsification. Once suspended the contaminants can be rinsed off using water. This is the most common form of decontamination in the field. Advantages are water and soap are inexpensive and readily available. Disadvantages are the material can retain its hazardous properties, runoff must be controlled and disposed of.

- **Evaporation** is the process of allowing the material to off gas. This is most often used when responding to gas releases such as ammonia or chlorine. It can also be used on materials with high vapor pressures. Advantage is it requires no special material or equipment. Disadvantage is it may take too long on some materials.

Chemical decontamination is the removal of hazardous substances by making them less harmful through chemical alteration:

- **Chemical degradation** is the process of using a chemical agent to chemically alter the contaminant. This process is almost never used on people. Advantage is it removes the hazardous properties of the substances. Disadvantages are you must use specific chemicals in the process, usually creates heat and sometimes toxic byproducts.

- **Disinfection** is the process of using antiseptic agents to kill microorganisms. Most commonly uses bleach on etiological agents. Advantage is disinfecting agents are readily available and inexpensive, can be done on site easily. Disadvantage is disinfecting agents may have dangerous properties and must be disposed of. Also, it is only effective on biological incidents.

- **Neutralization** is the process of turning an acid or base to a neutral pH. Advantage is it removes the dangerous properties of the material. Disadvantages are it cannot be done on people, can create heat and toxic gases, it takes a long time and requires large amounts of neutralizing agents.

- **Sterilization** is the process of killing microorganisms usually using heat, steam or radiation. Advantage it removes the hazardous properties of the material. Disadvantages are it cannot be done on people, it is a process done off site and is expensive.

- **Solidification** is the process of turning a liquid into a solid through a chemical reaction. Advantage is it almost eliminates vapor production and is easy to dispose of. Disadvantage is it requires specific solidification materials to perform.

Because most decontaminations are chemical specific, there are no set rules, but at a minimum, the firefighter should understand the procedures for emergency decontamination.

Gross decontamination is a quick fix to initially remove as much contamination as possible. Anyone contaminated at an incident needs to have the hazardous material removed as quickly as possible. Victims include those with dangerous contaminants on them who are in dire need of medical attention or responders who have been accidentally exposed to the product. This process occurs before technical or emergency decontamination can be formally established. For instance, an engine company arrives and finds contaminated victims or finds themselves unintentionally contaminated. Quickly deploying a hoseline or portable shower and conducting a gross rinse should remove a significant amount of the contaminant. It may also be necessary to remove contaminated clothing. Due to the urgent need of gross decontamination responders, many times will not deploy tarps or pools to catch runoff. Any delay in this could have serious consequences. Fire personnel have started to conduct a gross rinse of their structural firefighting gear after firefighting activities have concluded. This tends to remove most soot and debris collected on their gear and reduces the risk of exposure to the dangerous contaminants found in structure fires. Unfortunately, gross decontamination does not completely remove all contaminants from the victims and will require additional decontamination once a formal emergency decontamination line can be established or at a medical facility.

Emergency decontamination. Emergency decontamination is deployed at incidents that have ambulatory and/or non-ambulatory contaminated victims, responders that experience PPE failure or are having medical problems (fig. 4–15). As a minimum, the emergency decontamination area should have a tarp or catchall to protect the environment (fig. 4–16), a soap and water wash with a water rinse. For ambulatory victims, they should be instructed to remove contaminated clothing and step into a wash station or portable shower. Removing contaminated clothing will eliminate most of the contaminant but unless the victim's clothing is saturated it is often unnecessary to have them remove undergarments. Always attempt to provide for victim's privacy when removal of clothing is necessary. Victims personal belongings should be bagged and tagged. Once decontamination is completed they should be given a blanket or dry clothing and sent to medical for evaluation. For non-ambulatory victims, it will be necessary for protected responders to conduct victim rescue. The victim will be transferred to a backboard or stokes basket and have the contaminated clothing removed. Most ambulances won't haul the basket to the hospital, and the basket can hold product, depending on the type. They will be washed as thoroughly as possible, wrapped in a blanket and moved to medical for transport. Emergency decontamination is designed to be conducted under emergency conditions, fast to set up and conducted efficiently but quickly. Unfortunately, emergency decontamination may not remove all contaminants and a more thorough decontamination will need to be performed at the healthcare facility. During emergency decontamination, minimum protective clothing should include firefighting PPE, SCBA, and chemical protective gloves. In addition, depending on the substance, chemical protective clothing may be warranted. If responders become cross contaminated while performing emergency decontamination they will need to be decontaminated.

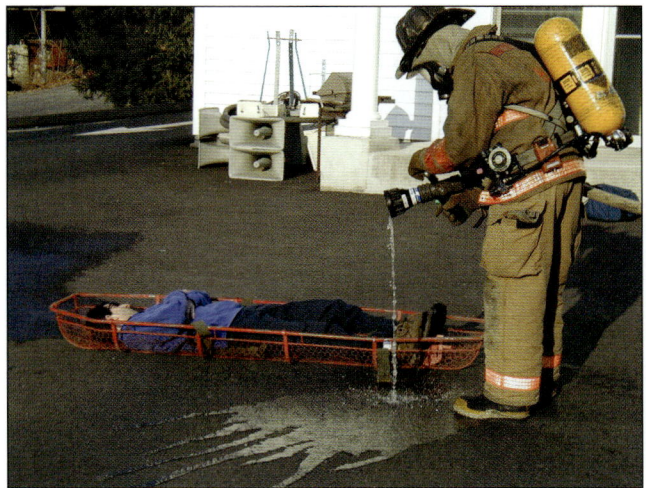

Fig. 4–15. Responder preparing to perform emergency decontamination on a patient.

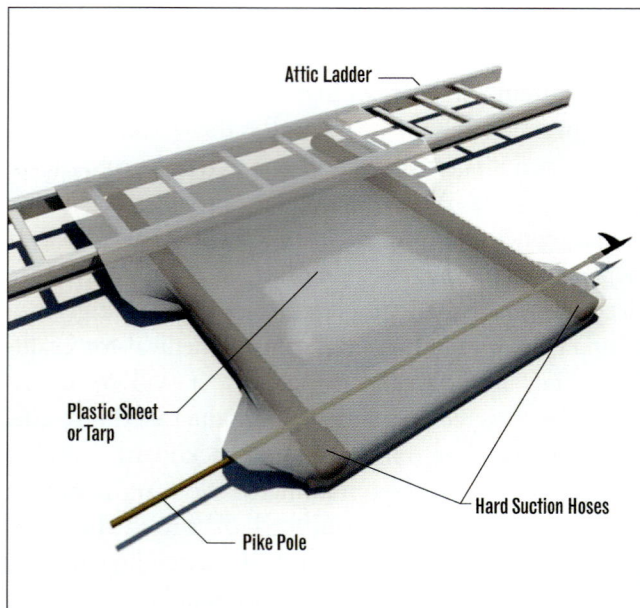

Fig. 4–16. A simple catch-all can be made with a ladder and a tarp.

The following procedures should be adhered to during the emergency decontamination process:

- Remove victim from contaminated area.

- Wash with flooding quantities of water.

- Remove all contaminated clothing.

- Continue to wash the victim.

- Move victim to an uncontaminated area.

- Transport to the hospital as soon as the victim is decontaminated.

- Advise ambulance and hospital personnel of the contaminant involved.

Some things to consider when performing emergency decontamination include the following:

- Avoid storm drains and any other areas where the environment could be compromised by the contaminants.

- If using safety showers in a facility, find out where the runoff is going. If it is going to the sanitary sewers, find an alternative.

- When removing chemicals that are corrosive in nature, use large amounts of water. Small amounts can cause a reaction with the corrosive.

- Make sure the flushing of water continues for at least 20 minutes when decontaminating corrosives.

Technical decontamination. HM 6.4 Technical decontamination is a more comprehensive approach to decontamination than gross or emergency decontamination and is used on protected responders. The most important part of the decontamination process is to limit the amount of contamination possible. If responders don't get dirty, they don't have an excessive need for decontamination. Properly trained and equipped hazmat responders should avoid puddles and not kneel down, crawl, lie in the product, or put themselves in a position to be showered with chemicals. When they are approaching the decontamination area, the less product on them, the less danger they are in. The more contaminated you are the greater the chance of exposure.

The necessary type, method and components of technical decontamination are dependent on the chemical and physical properties of the contaminant. For instance, if the incident involves a gas such as chlorine you may only need to use evaporation or a single gross rinse for decontamination. However, if the incident involves an oily or sticky substance then a gross rinse and multiple wash rinse station may be necessary. The command staff decides the type, method and necessary components included in the formal decontamination line. Components for a decontamination line may include:

- **Tool drop**: a location for the exiting entry team to leave tools so the new entry team can locate then.

- **Outer glove drop/hand wash**: a responder's hands and feet are usually the most contaminated areas. This allows the contaminated responders a place to quickly reduce the amount of contamination.

- A gross rinse station to remove the bulk of the loose contaminants (fig. 4–17). Because the entry person is the dirtiest he or she can be at this point, this stage should be solo. This solo wash can be accomplished by a shower, polyvinyl chloride (PVC) washing stalls, a portable stall, or a garden hose.

- **Outer wash and rinse** is an assisted wash and rinse using decontamination personnel (figs. 4–18 a and b). Depending on the contaminant a second wash rinse station may be required. At the end of the wash and rinse decontamination personnel must assess the level of decontamination. This is accomplished by visual inspection and using detection equipment such as pH paper. Decontamination personnel are usually dressed in one level below the entry personnel and will use the appropriate decontamination solution to wash and rinse entry personnel.

- **Undress** is a station where the CPE and SCBA is removed. The SCBA facepiece should be removed last. During this process, the entry personnel should only touch the inside of the suit and the decontamination personnel should only touch the outside of the suit. Even though contamination levels should be "as low as possible" different personnel should remove the SCBA.

- **Post entry medical exam**: on-scene medical examination and rehabilitation should be performed (fig. 4–19). An exam will compare vital signs taken before entry to vital signs taken after decontamination. If the entry involved the use of chemical-protective clothing, heat stress reduction and hydration of the individual should be a priority.

- **Personal shower**: full body wash, drying, and dressing of the individual. This may be performed on site or off site. Supplies such as shampoo, soap, scrub brushes, cotton swabs, gowns, and slippers must be made available to adequately carry out this stage

Fig. 4–18a. An assisted outer wash and rinse.

Fig. 4–17. A gross rinse is designed to remove the bulk of contaminants.

Fig. 4–18b. An additional wash and rinse.

Fig. 4–19. Final step in decontamination is a medical evaluation.

Record keeping and exposure records should be completed at this time, along with discussion of the future effects of the chemical. Post-exposure discussions should include what to look for if there is a delayed effect of exposure. Incident stress should be discussed. Many of these chemicals may be reproductive toxins or carcinogens. There may be psychological stress that needs to be diffused.

Site selection

Many factors go into selecting the best site for the decontamination corridor. The decontamination corridor is set up in the warm zone or the decontamination reduction zone. This zone separates the hot and cold zones. During set up of the decontamination line personnel are operating in a clean atmosphere. The decontamination line should be as long as necessary and as close to the hot zone as possible. The site should ideally be on a flat, hard surface set uphill and upwind. It should be as close to a water supply as possible and have access to electricity and lighting. Ensure there is adequate runoff control and sewers and drainage should be blocked as a precautionary measure.

Decontamination considerations

Weather is a major factor in hazardous materials response. In extremely hot temperatures it is helpful to provide decontamination personnel a shelter to stay out of the sun while waiting for the entry team to come through the decontamination line. Also, the team leader and safety officer should ensure personnel hydrated prior to suiting up. After completing the assignment decontamination team members should be evaluated for signs of heat stress. In extreme cold weather, indoor sites are preferred or possibly use tents at the very least attempts should be made to place barriers up to block the wind. Consider using dry methods of decontamination if possible.

Because decontamination is so equipment- and personnel-intensive, it is imperative that it be practical and practiced. Planning, contingencies, proper equipment, and training will prove to be the difference between failure and success. A hazardous material incident is no place to rehearse. All responders need to be familiar with the local emergency response plan and department SOPs as they relate to the technical decontamination process.

Mass decontamination

HM 6.3 **Mass decontamination** is the removal of contaminants from a large number of contaminated victims who are typically in life threatening situations. Establishing mass decontamination is necessary when the number of victims overwhelms normal decontamination capabilities of local responders. It is preferred to control the runoff, however, due to the chaotic life-threatening situation and the significant amount of water being used, controlling runoff is extremely difficult and often times the practice is abandoned. The circumstances where mass decontamination is necessary would be disseminating chemical or biological agents at places where there are a large number people gathered such as a sporting event, fair or political gathering. While there is a chance that the cause of the release is accidental such as a train derailment near the event or gathering it is much more likely this is caused by a deliberate act. Which often makes the incident a criminal act adding to the complexity of the incident. While response agencies will not know the act is planned they will know they event is planned. Many communities hold extensive meetings planning large events addressing crowd control, traffic routes, additional staffing for law enforcement, fire and ems needs, they should also plan for mass decontamination. There is very little time to implement mass decontamination if it is to be effective. Therefore, the time to design the plan is not after the dissemination of the chemical agent. There are many factors that must be addressed to effectively implement mass decontamination and they are as follows:

- **Personnel.** Adequate fire, police and EMS numbers to provide for victim rescue, crowd control, decontamination set up/implementation, medical treatment/transport, evidence collection and scene security.

- **Location.** Selection of a location to establish the decontamination line.

- **Apparatus.** Adequate number of fire apparatus and ambulances.

- **Water supply.** Decontamination is set up close enough to an adequate water supply.

- **Clothing and valuables collection.** Often time it is necessary to have victim remove contaminated clothing. There must also be a method to collect, bag, tag and record valuables.

- **Victim privacy.** It is always preferred to string tarps up to provide for victim privacy. If possible have two lines for ambulatory victims, one for each gender. If possible have female responders assisting female victims and male responders assisting male victims. Always allow children to be accompanied by at least one of their parents.

- **Inside facilities.** In the event of extreme cold temperatures responders should consider arranging for an inside area such as a school locker room and showers.

- **Alternative clothing.** Provide light weight Tyvek suits for the victims to change into, a location to change and be provided medical triage and transport. This can be done at a school gymnasium or convention center.

Implementing mass decontamination. Based on the individual staffing positions in the action plan responders should don the appropriate PPE. All responders that perform victim rescue or are working in the decontamination line should be in structural firefighting gear and SCBA as a minimum. Generally, mass decontamination is accomplished by positioning fire apparatus and hoselines to provide a gross wash to contaminated victims (fig. 4–20). There should be a place to collect contaminated clothing for isolation and disposal and a bag and tag system for valuables. Once the mass decontamination line has been established responders must communicate clearly with ambulatory victims. Victims should be directed to an area of refuge where they will be prioritized and directed pass through the decontamination line then to medical triage, treatment and transport. Victim priority is based on those who are exhibiting the most serious signs and symptoms to no signs and symptoms or those who are the most contaminated to the least contaminated. There is a distinct possibility that some of the victims will be unable to walk due to illness

or injury. These victims will need to be rescued from the hot zone and a separate decontamination area should be established for these nonambulatory victims. The separate decontamination area is to allow the ambulatory victims to pass through as quickly as possible. This will put an additional strain on staffing requirements so requesting mutual aid should happen as early as possible. Many times, deceased victims will be left for the medical examiner to evaluate. However, it is still necessary to decontaminate those once all injured victims have been processed.

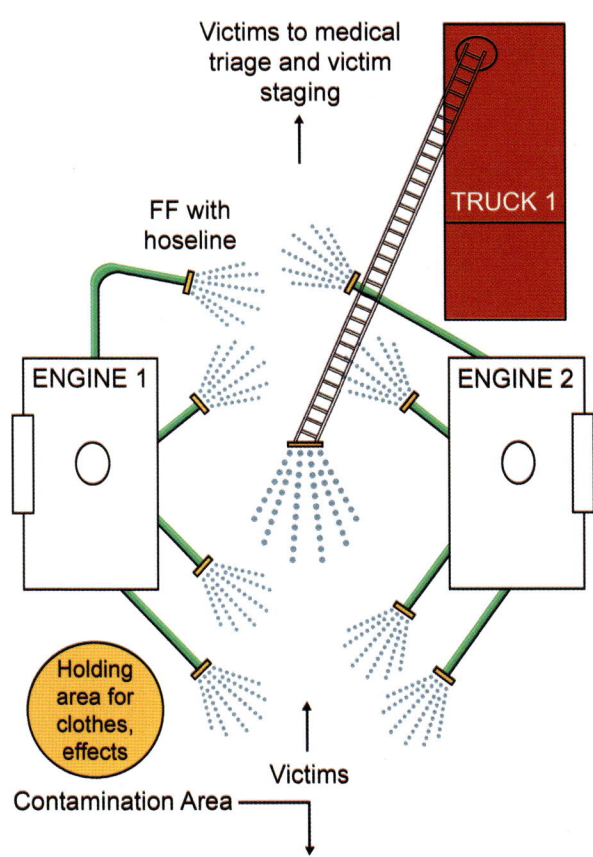

BASIC PLAN
Hazardous Materials Response Team
Mass Casualty Decontamination

Fig. 4–20a. Examples of a basic a mass decontamination plan (after FEMA and Baltimore County [MD] FD)

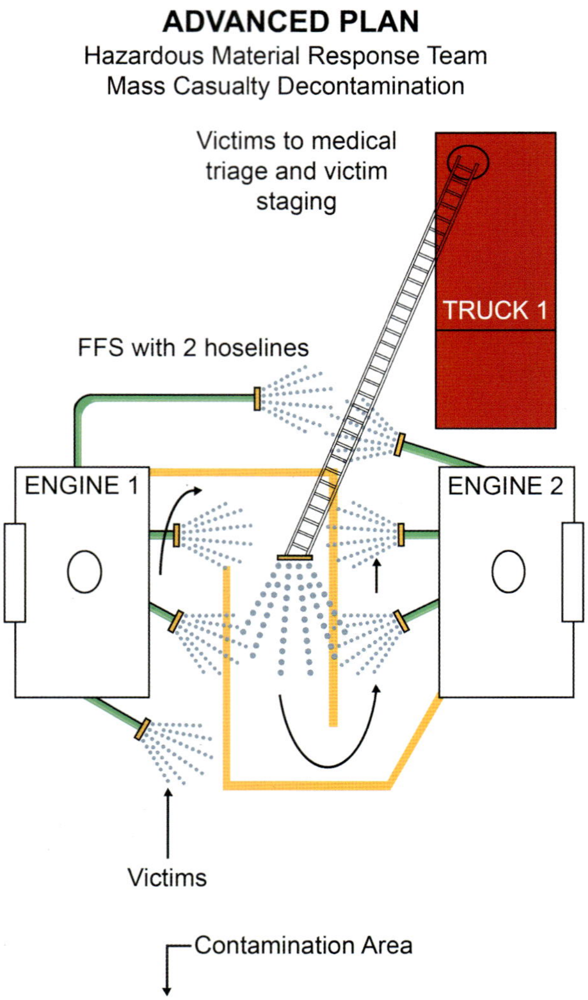

ADVANCED PLAN
Hazardous Material Response Team
Mass Casualty Decontamination

Victims to medical triage and victim staging

TRUCK 1

FFS with 2 hoselines

ENGINE 1

ENGINE 2

Victims

Contamination Area

Fig. 4–20b. Examples of an advanced mass decontamination plan (courtesy of FEMA and the Baltimore County, MD, Fire Department)

EVIDENCE PRESERVATION AND PUBLIC SAFETY SAMPLING

Crime scenes

HM 6.5 Not all hazardous materials incidents are accidental releases. Sometimes hazardous materials releases involve criminal activity such as terror attacks, illegal dumping, and illicit labs (fig. 4–21). All of these present unique hazards to responders and must be treated as crimes scenes. In addition to the standard hazardous materials response objectives, responders must collect and preserve evidence. All personnel operating at the incident will be required to wear appropriate PPE based on the hazards identified. There is also an increased presence of law enforcement to process the crime scene. Some jurisdictions train law enforcement personnel to don and work in CPE so they can better process the scene others provide additional training to firefighters to collect and preserve evidence at intentional hazardous materials incidents. Operations level personnel trained to the mission specific operation of evidence preservation and public safety sampling must operate under the guidance of a hazardous materials technician, an allied professional including law enforcement or others with similar authority, an emergency response plan or standard operating guideline.

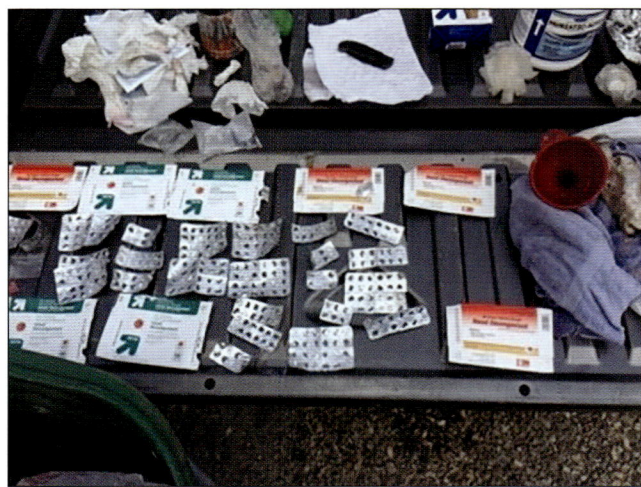

Fig. 4–21. Example of hazardous materials incident that is a crime scene. Note the packets of pills and pharmaceutical boxes.

The hazards present at intentional incidents are much the same as accidental releases such as flammability, corrosivity, oxidizers and responders are required to wear the necessary PPE as approved by the command staff. This will range from structural firefighting gear to chemical protective clothing. Respiratory protection can range from APRs to SCBAs. Also, there can be additional hazards at intentional releases such as secondary devices designed to target first responders, booby traps at illicit labs designed to target intruders that includes first responders and some crime scenes will have armed persons. This will often require law enforcement to wear body armor and members of the bomb squad to wear blast suits. All of these additional hazards must be dealt with before hazardous materials response personnel can make entry to take care of the incident.

Recognizing crime scenes. Not all intentional releases are reported as such to 911. Fire and EMS personnel must be cognizant of the signs indicative of illegal activities. Many times, illicit labs are involved in fires or those operating the lab are overcome with chemicals used and are found unconscious. If responders find lab equipment such as beakers, flasks glass tubing or burners there is a good chance they have happened upon a drug lab, or petri dishes may be someone growing biologicals. Incidents that have multiple patients exhibiting the same signs and symptoms or there are garden sprayers found in odd places it could be suspected that a chemical or biological attack has occurred. Many times suspicious packages or letters are delivered to government buildings, controversial facilities such as abortion clinics or animal testing laboratories, celebrities of government officials. Signs of suspicious packages or letters include:

- Excessive postage

- Oily stains, crystallization or discolorations on package

- Excessive weight

- No return address

- Rigid, lopsided or uneven envelopes

Once fire and/or EMS personnel suspect that they could be at an intentional release they should immediately notify law enforcement. Also, due to the possible chemical, biological or explosive hazards that may be present they should retreat to a safe area until a proper hazard analysis can be completed. The only exception to this could be if they are conducting a rescue. Once the rescue is complete the incident should be secured and personnel evacuated. Depending on the nature of the incident will determine who is notified for investigative purposes. Initially, local law enforcement should be notified and they will be responsible for crime scene management. Crime scene management can include rendering the incident safe, scene processing, evidence collection and preservation as well as notification of state and federal agencies (fig. 4–22). Many different agencies are authorized to investigate criminal incidents. Some illicit labs are handled by local law enforcement while others, such as large drug labs may include state law enforcement or the DEA. Environmental crimes should include the state health department and the EPA, terrorism is controlled by the FBI and explosive incidents may bring in ATF. Once the incident has been declared a crime scene all state and federal notification should be done by local law enforcement.

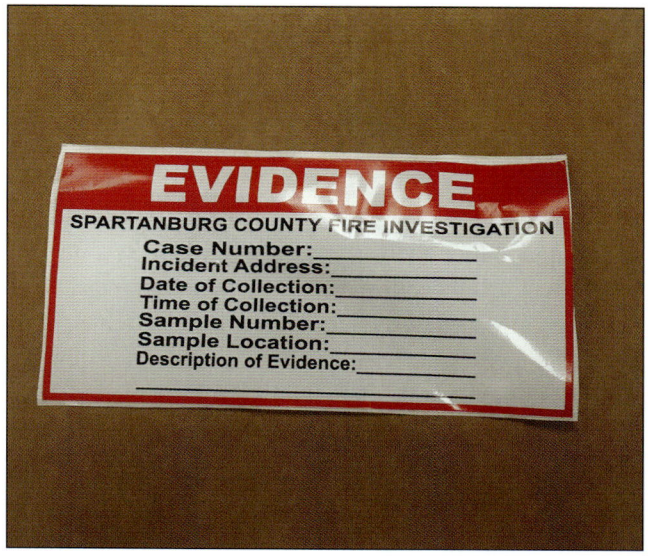

Fig. 4–22. Evidence collection is part of crime scene management

Response objectives for crime scenes. Once a hazardous materials incident has been declared a crime scene it adds a new and exciting dynamic to it. Due to the increased complexity, incident command switches to a unified command post. Unified command is a system where a representative from each agency is present in the command post and command responsibility is shared depending on which role is in the lead.

Initially law enforcement takes the lead to render the scene safe from hostile attack or explosives. Once that is accomplished, life safety and hazard mitigation should take the lead. This includes rescue, treatment and transport of injured victims, evacuation/shelter in place, mitigating the hazardous materials and providing decontamination. During hazard mitigation, hazardous materials works should take extra care to keep the scene as undisturbed as possible or take photographs. When all victims are taken care of the hazard is mitigated crime scene processing begins. Hazard mitigation means that it is no longer a threat to the public. It doesn't mean that chemical protective clothing is no longer necessary. That doesn't happen until a contractor has completed the cleanup. This is the time when the decision is made to have law enforcement personnel process the evidence and crime scene in CPC or have hazardous materials personnel collect and preserve evidence (fig. 4–22).

Either way law enforcement is responsible that all necessary search warrants are obtained. If the evidence gets contaminated or a proper chain of custody isn't kept valuable evidence may be found inadmissible in court so it is imperative the scene is processed correctly. This begins with scene security. Besides the normal hot war and cold zones, law enforcement must establish a crime scene perimeter and document everyone that has entered the crime scene.

Collecting and preserving evidence.

The first thing that has be understood is that all evidence must be accounted for at all times. This is accomplished by tracking evidence from the time it is collected until it is secured in law enforcement evidence storage. This practice is known as chain of custody. Each agency has a chain of custody policy and it must be followed. When the evidence is discovered and packaged a chain of custody for is filled out. That form must follow the evidence until it is disposed of or it is returned. The chain of custody form records the name of the person taking custody along with the date and time of transfer, sample ID number, description, and the evidence location (fig. 4–23). If there is a gap in chain of custody that evidence may no longer be viable for court.

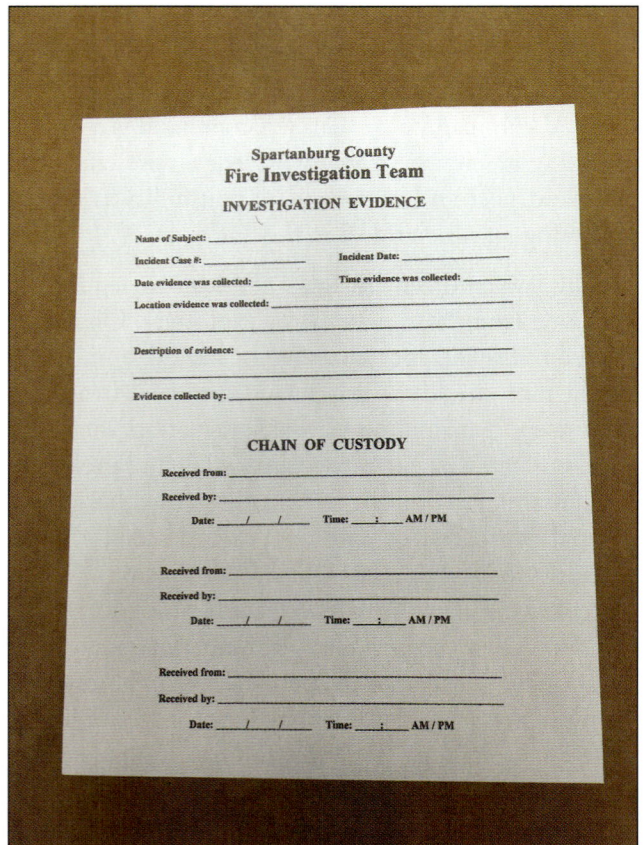

Fig. 4–23. Chain of custody form

Crime scene processing is conducted through a scientific method where investigators will develop a hypothesis and work the scene trying to prove it. Evidence identification should be accomplished by qualified investigators because to the untrained eye many pieces of evidence will be overlooked. During hazard mitigation, a hazardous materials responder may find it necessary to collect or move potential evidence in order to preserve it. If this occurs it is crucial that any actions taken be meticulously documented. Personnel involved in any evidence collection or preservation is subject to testify in court.

Public safety sampling.

Hazardous materials response personnel may be asked to collect samples of the hazardous materials as evidence. After collection, the sample must stay with the entry team through decontamination and then get transferred to law enforcement. The responders who perform public safety sampling are part of the chain of custody and subject to being subpoenaed.

While public safety sampling is one of the most critical aspects of the investigation, it only occurs after victims have been taken care of and the incident has been cleared of explosives, booby traps or armed assailants. Materials used in criminal acts are not usually labeled, therefore one purpose for sampling is to perform an analysis on the material to assess its hazards, attempt to identify the material, and mitigate the incident. The remaining samples are kept for evidence. Therefore, responders must try to collect enough material to provide for testing and evidence. Also, all samples must be obtained and accurately recorded in accordance with law enforcement evidence collection policies.

Once the material(s) have been located responders will attempt to identify hazards at the site. This is not an identification of the material but a hazard classification of the material such as radioactive, flammable or corrosive. With the hazards identified, hazardous materials responders can provide for victims, ensure decontamination is adequate and the site can be processed in a relatively safe environment. Hazard classification is usually completed on site through monitoring strategies. During this phase, hazardous materials responders are in appropriate PPE and use radiation detectors, air monitoring equipment and chemical/biological detection papers or tubes. All air monitoring readings must be recorded and all detection papers should be labeled, catalogued and sealed as evidence. After this, samples are collected, labeled, documented and transferred to law enforcement. The samples will be sent to a lab for positive identification and then placed into evidence. Due to the Laboratory Response Network (LRN) many local testing laboratories have been upgraded to provide state of the art testing of chemical and biological samples. The LRN was established by the CDC in 1999. Since, it has expanded to include many local and state run labs. All labs operated under the LRN require the use of approved labeling, packaging, documentation and transportation. Photographs or videos of the site should be acquired prior to and during sampling to document the site and samples.

Obtaining samples. All sampling equipment must be clean prior to use. This means that if you have multiple materials or are taking samples from multiple locations a fresh set of equipment must be used for each. For collecting evidence, special sampling equipment and containers may be required. These items are sealed and certified as clean. Local law enforcement policies will dictate the types of equipment and containers necessary for obtaining samples. Sampling equipment includes (fig. 4–24):

- Jars
- Bottles
- Sealable bags
- Pipettes
- Coliwasa tubes
- Scoops
- Shovels

Fig. 4–24. Sampling equipment

Sampling for airborne contaminants responders will use air monitoring equipment and ensure the results are properly documented.

Sampling is conducted in teams of at least two members. One member is responsible for the container and the other responsible for the sampling equipment. At no time should the container or equipment touch a contaminated surface. If this occurs the equipment must be discarded and a fresh set used.

Containers should be numbered prior to entering the hot zone. The collection team should note the condition of the collection site. This can be accomplished using a note pad, with photographs or video. Using the sampling equipment, obtain a sample and place it into the numbered container, label and seal it. The sampling team should also bag any physical evidence found at the scene such as sprayers, boxes or bags that held the contaminant, letters or any out of place items. These items should be double bagged to prevent spoilage of forensic evidence such as DNA or fingerprints. It is also common to bag the sample container as a secondary protection of the sample. The sampling team must take all sealed and labeled samples through decontamination and have the exterior of the sample containers decontaminated. To maintain chain of custody the sampling team should remain in possession of the containers throughout the decontamination process or at least in sight of the containers.

After decontamination, the sampling team must deliver all samples to the law enforcement personnel responsible for evidence collection. Labels are verified and chain of custody forms are completed before the sampling team is permitted to part with the samples. If available, field testing of samples can be conducted using reagent papers, air monitoring, radiological detectors, PID's and fluoride tests (fig 4–25). Testing should be conducted for radiation, corrosivity, flammability, oxidizers VOCs and fluoride. Personnel performing the tests should wear appropriate protective clothing and safety precautions. Once the field tests are completed samples should be sealed and sent to the lab. Transportation to the receiving lab is the responsibility of the law enforcement agency having investigative authority. Keep in mind, local, state and federal law enforcement has different sampling and labeling policies.

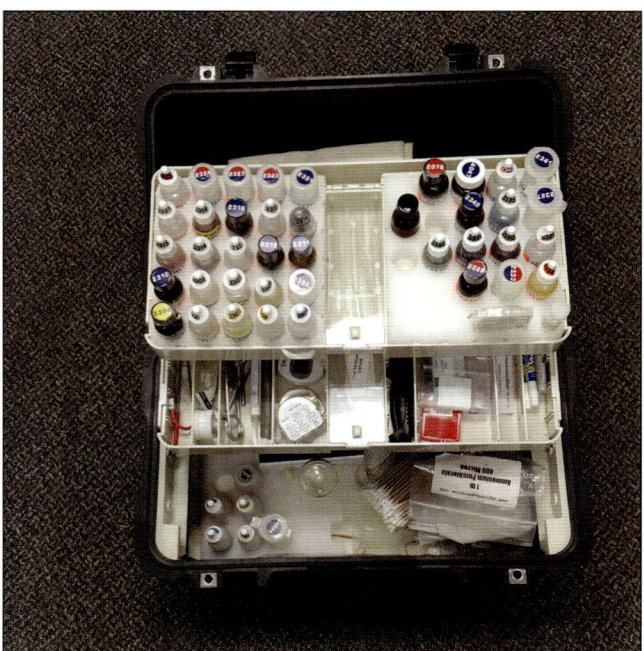

Fig. 4–25. Field testing equipment

PRODUCT CONTROL (DEFENSIVE OPERATIONS)

Objectives

HM 6.6 Upon completion of this section, you should be able to do the following:

- Explain the types of spill control.
- Explain the types of leak control.
- Explain the types of fire control.

The efforts of firefighters to contain the release of a hazardous material are intended to minimize environmental effects by confining the spill to the smallest possible area for later removal by commercial cleanup contractors. As with all other activities during these types of incidents, the safety of firefighters is vital, and those performing these tasks must make every effort to avoid contact with any hazardous material involved.

Hazardous materials can migrate and enlarge an incident scene. Spilled liquids move downhill where the product can then enter storm drains or other bodies of water; vapors from spilled liquids and dust from finely ground spilled solids move by air currents. Remember, anything used to confine a hazardous material must be compatible, meaning that the confinement barrier cannot react with the hazardous material, which could produce

dangerous fumes or perhaps even a fire. Another word of caution: avoid any contact with spilled hazardous materials during confinement efforts.

Product control options

Defensive control methods are used to contain or confine a spill. These actions should be undertaken only if it is possible to perform the tasks without contacting the material. All responders must be wearing the appropriate PPE for the hazard. Responders need to be able to identify the purpose for and the procedures, equipment, and safety precautions associated with each of the following control techniques.

Absorption. Absorption is the physical process of absorbing or "picking up" a liquid product spill (fig. 4–26). In this process the liquid enters the interior of the absorbent material used and is trapped in that product. It is most effective on spills less than 55 gal (208 L). Some absorbent materials can be used for liquid spills on both land and water. Booms and pads may be used to absorb materials on the surface of water. The absorbent material should be spread on top of the product or in the direction of its travel.

Fig. 4–26. Using absorbent materials to "pick up" a fluid spill

The following materials can be used for absorption:

- Dirt
- Sawdust
- Absorbent pads
- Socks
- Booms
- Speedy dry

- Absorbent particulate
- Pans
- Pillows
- Charcoal
- Kitty litter
- Product-specific absorbents

The first responder should consider the compatibility of the absorbent with the spilled product. For example, spreading sawdust on an oxidizer could result in a fire. The properties of different absorbent materials vary, and they are not necessarily interchangeable.

It is important to understand what types of liquids will be absorbed by the absorbent material being used. Some will not pick up water, whereas others will pick up any liquid they come in contact with. This information is helpful when determining what is needed to soak up an oil spill in a lake or stream. Because it is considered hazardous waste, the first responder must ensure that the absorbent material and the product is disposed of in accordance with environmental regulations.

The absorption method should not be confused with adsorption, which is when one material adheres to the outside of another.

Adsorption. **Adsorption** is the chemical process by which the spilled product adheres to the surface of the sorbent. Most adsorbents are manufactured and the most common adsorbent is activated charcoal. As with absorbent the absorbent material must be compatible with the spilled product.

Confinement—diking. **Diking** is the process of constructing a land-based barrier that controls the movement of liquid hazardous material. Dikes are used to stop the spread of the hazard zone and prevent movement of materials into areas where they can cause greater damage. They are constructed by forming an embankment. Construction should begin at the point farthest away from the spill and work back to the spill (fig. 4–27). The responder should ensure that it is far enough away from the spill that it can be completed before the material reaches it. Dikes can be constructed of materials available on site such as dirt, sand, clay, and other materials that may be found on the incident scene. For small spills there are two types of dikes commonly used to capture and control product releases:

The V-dike captures released material by using the small end of the V as the collection point.

Fig. 4—27. A circle dike

The circle dike is used as a secondary containment to capture a released material that escapes or overflows the V-dike. It is a berm that is placed 360 degrees around a leaking container. It does not form a perfect circle but rather an enclosure for the product.

For larger spills such as an overturned tanker will require the use of a larger dike that is slightly angled across the path of flow. Sometimes multiple dikes are necessary to confine larger spills.

Confinement—damming. **Damming** is the process of constructing a barrier intended to slow or stop the flow of liquid into the environment. A dam or line dam can be constructed of any materials available that could prevent a liquid from moving down an incline. They are typically built in ditches, streams, or creeks. Additionally, dams may be used to act as accumulation points where product can be collected for cleanup. Overflow and underflow dams are used when constructing a dam in moving water.

Overflow damming. **Overflow dams** should be constructed from materials with a specific gravity greater than 1. Hazardous materials with a specific gravity greater than 1 will sink in water and will be trapped against the base of the overflow dam. The pipes at the top of the dam allow the water that remains on top to flow freely over the dam (fig. 4–28). Pipes used in overflow dams should be a minimum of 4 in. in diameter.

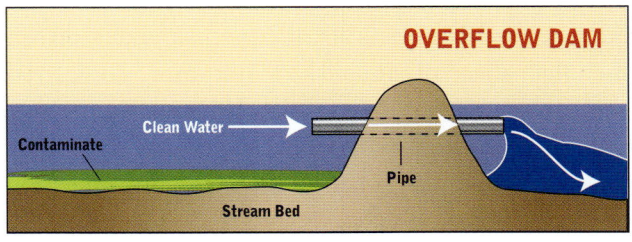

Fig. 4—28. An overflow dam

Hazardous Materials Operations: Mission Specific

HAZMAT

Chapter 4

Underflow damming. **Underflow dams** should be used for materials with a specific gravity lower than 1. Hazardous materials that have a specific gravity lower than 1 will float on water and will be trapped against the top of the overflow dam. The pipes at the bottom of the dam allow the water at the base of the dam to flow freely (fig. 4–29). Because most spilled hazardous materials float on water, this type of dam is most commonly used.

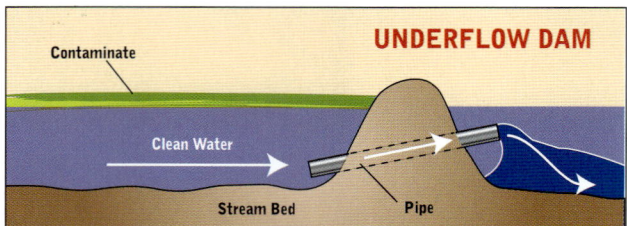

Fig. 4–29. An underflow dam

The following materials may be used to construct dams:

- Dirt or sand

- Pipes

When using pipes, note that they should have at least a 4-in. (100-mm) diameter. If the water flow is larger, it may be necessary to use either a larger pipe or multiple pipes. The responder should place enough pipes to cover two-thirds of the waterway. Hard suction can be used in an emergency.

Confinement—diverting. **Diverting** is channeling spilled materials into a containment area to an area where it poses less harm. It can be used on both land and water. A spill on land can be diverted by building a barrier made out of a material such as dirt, ahead of the spill. A spill on water can be diverted using booms to channel the material into an area where it can be absorbed or collected. Diverting allows the product to be directed to an area where it will cause less harm. For example, it may prevent a product from flowing down a storm drain (fig. 4–30).

The first responder should consider the angle and speed of the flow when constructing a diversion barrier. When dealing with fast moving spills, the barrier should be set up at an angle of no less than 60 degrees to be effective. As with dikes and dams, a diversion barrier should be placed far enough ahead of the spill so that it can be completed prior to the arrival of the material. This may require that some intermediate area between the spill and the barrier be sacrificed. Diversion barrier are generally built using the same materials as dikes.

Fig. 4–30. Spills can be diverted around an endangered area such as a storm drain using a diversion barrier.

Confinement—retention. **Retention** or **containment** can be broken down into two methods based on whether the physical state of the product is liquid or solid.

Liquid. When containing a liquid spill into water, booms are most commonly used. Booms are long tubular devices that float on top of the water and provide either a barrier or barrier/absorption (fig. 4–31).

Fig. 4–31. Multiple absorbent booms

Absorbent booms are made from various materials that can collect compatible materials released during a spill.

Retention or barrier booms are constructed of nonporous materials that float on the water and have a curtain extending below the surface to hold the product in place. The product is then removed using either absorbent pads or other manual devices, or it is vacuumed up by a cleanup contractor.

Solids. When containing spills of solid materials, the method most often used is covering the product. This involves placing tarps over the material to prevent movement. Movement may be caused by weather conditions such as wind and rain or by personnel walking through the product.

Dilution. Dilution reduces the concentration of the material to a nonhazardous or less hazardous state (fig. 4–32). This method has few practical applications at a hazmat incident, especially for first responders. This process can be used when dealing with small acid or base spills. There are disadvantages to using this method. Dilution is more likely to increase the volume of the product and create a runoff problem. It may also weaken the original product but not eliminate the hazard. In addition, fuels, oils, and other hydrocarbons are not water-soluble and cannot be diluted with water. Finally, this method should not be used on water-reactive materials.

Fig. 4–33. Firefighters dispersing a propane vapor cloud

When using this technique, first responders should confine the runoff and analyze it for contamination. When applying this technique to flammable vapors, such as LP gases, be aware that reducing the concentration in air may bring it into its flammable rage. Vapor dispersion is usually used to remove gases that are heavier than air from low-lying or enclosed areas. It is usually not a recommended practice except in life-threatening situations.

The first responder must know the identity of the product before using this tactic. Lighter-than-air vapors, which would normally dissipate on their own, may be knocked down by water spray.

Water-soluble materials such as anhydrous ammonia, when mixed with water, may produce another substance that could be dangerous. For example, anhydrous ammonia when mixed with water produces ammonium hydroxide, which is a corrosive liquid.

Fig. 4–32. Dilution reduces the concentration of the material to a nonhazardous or less hazardous state.

Vapor dispersion. By moving gases or vapors using **vapor dispersion**, the hazard and concentration of the product are reduced, while firefighters are able to work from a safe distance. Vapor dispersion uses water spray to direct hazardous vapors away from a certain area (fig. 4–33). The turbulence created by the stream mixes up the air and reduces the material's concentration. Fans may also be used to prevent fewer problems with runoff and chemical incompatibility with water. However, fans should not be used with flammable materials unless a unit such as a water-driven positive-pressure fan can be used to reduce the risk of ignition. To be effective, the material must be water soluble or the vapor cloud must be moveable. Be sure to disperse the vapors into an area where they will not cause more harm.

Vapor suppression. Vapor suppression is the reduction or elimination of vapors produced by a spilled hazardous material. Firefighting foams are effective when used on flammable or combustible liquids (fig. 4–34). This method significantly reduces the hazards associated with uncontrolled vapors. The selection of the proper foam for vapor suppression is an important decision. Transportation and industrial accidents, or even a mishap at the local high school, can result in a flammable liquid or hazardous vapor incident.

Fig. 4–34. Foam being applied to a combustible liquid

Foam extinguishes flammable and combustible liquid fires and protects liquid spills from ignition in four ways:

- Removes the air from the flammable vapors

- Minimizes or eliminates vapor release from fuel surface

- Provides a barrier between the flame and the fuel surface

- Cools the fuel surface and surrounding surfaces

The following are common types of foams used in hazardous materials applications:

- Aqueous film-forming foam (AFFF)

- Alcohol resistant aqueous film-forming foam

- Fluoroprotein

- High-expansion foam

The following factors must be considered prior to using foam:

- The product must be compatible with the type of foam being used (water-reactive chemicals cause adverse reactions).

- Personnel should not walk through or disturb the foam blanket after it has been applied and established.

- After the foam is applied, monitor the integrity of the foam blanket.

***Types of foam.* Aqueous film-forming foam (AFFF).** AFFF forms a thin layer of foam that floats on the surface of the fuel (fig. 4–35). It is a synthetically produced, detergent-based foam used on hydrocarbon

fuels. There are several advantages to using AFFF:

- Extremely rapid knockdown

- Long shelf life

- Easy to foam and can be used with nonaspirating nozzles

- Self-healing foam blanket

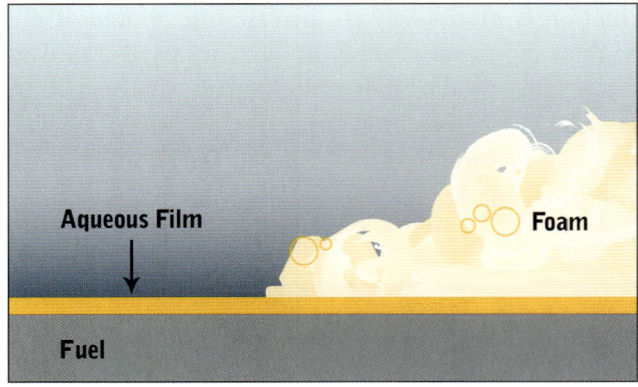

Fig. 4–35. Aqueous film-forming foam (AFFF)

Among the disadvantages of AFFF include:

- Less overall burn-back resistance than fluoroprotein foam

- Ineffective on polar solvents

- Effectiveness depends on the proper formation of an aqueous film

AFFF is self-healing after being disrupted and has good viscosity at low temperatures. It can also be purchased in an alcohol-resistant formula for use on alcohols and polar solvents.

When encountered with a polar solvent (alcohols, ketones, etc.) normal AFFF will be ineffective due to polar solvents ability to mix readily with water. Polar solvents require the use of alcohol resistant AFFF to produce an effective foam blanket.

Fluoroprotein foams. Although fluoroprotein foams are designed for hydrocarbon fires, protein foams can be formulated to be alcohol resistant by adding an additional mixture of materials. However, they generally only maintain their alcohol-resistant properties for about 15 minutes. Fluoroprotein foam can be injected into the base of a burning storage tank and be allowed to rise to the surface and extinguish the fire. This procedure is known as subsurface injection.

Fluoroprotein foam has the following advantages:

- Reasonably compatible with dry chemical extinguishing agents

- Better burn-back resistance than AFFF

- Not dependent on film formation

- Good fuel-shedding properties

- Nontoxic and biodegradable

Fluoroprotein foam has the following disadvantages:

- Does not provide as quick a knockdown as AFFF

- Requires aeration

- Not film-forming

This foam type is not affected by freezing and thawing and can be mixed with either fresh or salt water. In addition, fluoroprotein foams can be stored in temperatures ranging from 20°F (−7°C) to 120°F (49°C).

High-expansion foams. High-expansion foams are special-purpose foams with a detergent base. They are characterized by low water content and have poor heat resistance.

This type of foam is used for:

- Pesticide fires

- Suppression of fuming acid vapors

- Firefighting operations in confined spaces

Using high-expansion foam as a vapor suppression tool offers the following advantages:

- Good for use in basement fires or enclosed areas

- Low proportioning rate in conjunction with high-expansion ratios allow for large quantities of foam to be produced with a minimum amount of concentrate

- Not affected by freezing and thawing

High-expansion foams have the following disadvantages:

- Limited use on Class B fires because the thermal updraft causes difficulties in the formation of a thick blanket

- Does not heal itself when disrupted

- Generally not recommended for the outdoors because the foam blanket tends to blow off easily in the wind

Foam production. Foam is a specific agent for a specific type of material. As such, it must be used properly. Proportioning or mixing the right amount of foam concentrate with the right amount of water and air at the right pressure results in good quality foam. Foam is not just a collection of bubbles; it is a specific mix of concentrate, water, and air. It is also critical that there are sufficient quantities of foam concentrate to extinguish the fire and secure the spill.

Proper proportioning is important. If there is not enough concentrate in the finished foam, the overall quality and effectiveness will decrease. It may not extinguish the fire as expected, heat resistance decreases, and burnback resistance decreases.

If your finished foam has too much concentrate in it, foam will not flow around debris and obstacles. Additionally, limited foam concentrate will be depleted rapidly, not allowing enough for complete extinguishment and reapplication as necessary.

A common type of foam proportioning and delivery system is called an **inline eductor** (fig. 4–36). The eductor uses the venturi principle to pull foam concentrate from the foam container and mix it with the appropriate amount of water to make foam solution. Inside the eductor there is a restriction in the metering orifice that creates a **venturi effect**, causing the water to flow faster and creating a lower pressure area in the line, resulting in suction (fig. 4–37). The foam pickup tube is attached to the venturi area of the eductor and goes in the foam bucket. A metering device allows the passage of the correct amount of foam concentrate to enter the hoseline. Determining the appropriate ratio of foam to water depends on the foam being used and the material it is used on. For example, an alcohol-resistant AFFF with a 1%/3% application ratio (fig. 4–38) would require 1% of foam concentrate and 99% water on a hydrocarbon spill and 3% of concentrate and 97% water on an alcohol spill.

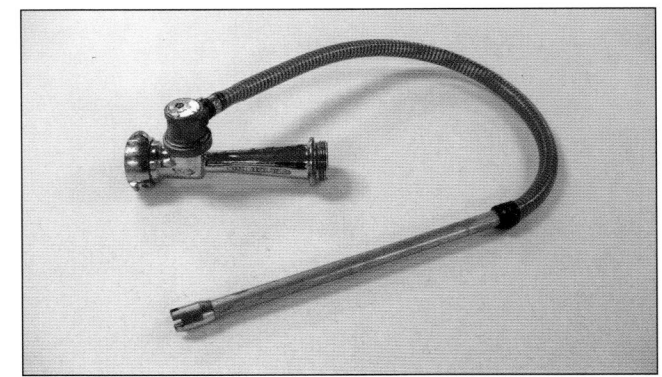

Fig. 4–36. Inline foam eductor

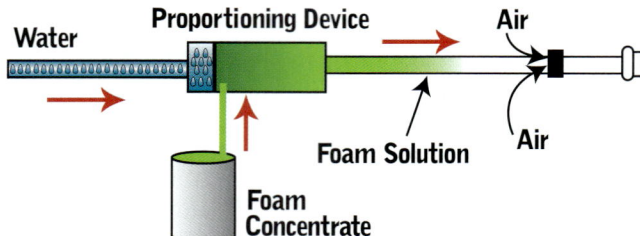

Fig. 4–37. The foam process introduces a set percentage of foam concentrate into the water supply and uses a foam nozzle to inject air into the foam solution.

Fig. 4–38. Alcohol resistant AFFF (courtesy of National Foam)

The nozzle is the last step in the foam making process. Here the foam solution is aspirated, and finished foam is produced. The air is entrained in the foam by some type of mechanical means, usually directed at a baffle, or by foam solution streams directed at each other in the nozzle, then directed down the foam tube and out the end, forming the foam stream (fig. 4–39). The best quality foam blanket is generated with a nozzle designed for foam making and distribution.

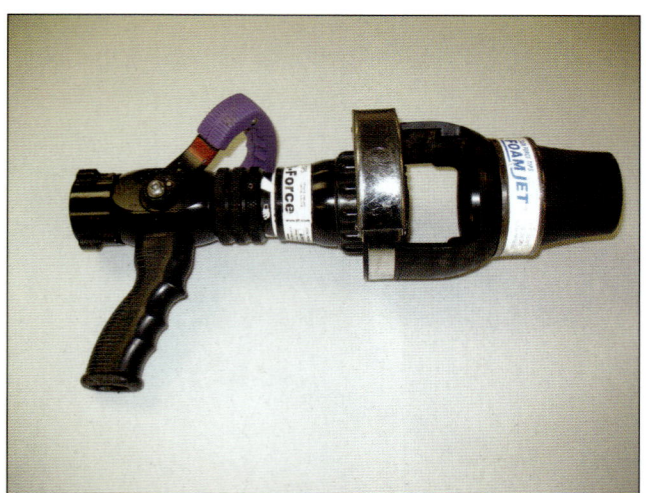

Fig. 4–39. Foam nozzle

Application of foam. It is the nozzle operator's responsibility to apply the foam as gently as possible to allow the finished foam to form a cohesive but flowing blanket on the fuel to cover the product, allow the best development of the film-forming actions (stop vapor production), and maintain the security to prevent ignition. When applying foam, the operations-level responder should be familiar with local department SOPs on foam equipment and application methods. Different types of foam use different methods of application. The bank down, roll-on, and rainfall techniques are methods of applying foam.

Bank-down technique. In this technique, foam is bounced off a wall or object behind the spilled product. The foam then falls onto the hazardous material (fig. 4–40).

Fig. 4–40. Bank-down technique

Roll-on technique. In this technique, foam is applied ahead of the spilled product and gently rolled over it (fig. 4–41).

Fig. 4–41. Roll-on technique

Rainfall technique. In this technique, the foam is sprayed into the air above the product and allowed to rain down upon it (fig. 4–42).

Fig. 4–42. Rainfall technique

Remote valve shutoffs. This technique is not typically used by first responders at the operations level, but it is the best method of controlling spills for a limited number of situations. This process requires the firefighter, either manually or through engineering control methods (hydraulic, pneumatic, or mechanical), to stop the physical flow of product from a valve. This is done from a remote area, not at the leak. This should only be attempted if the responder will not be placed in the hot zone and there is no danger of coming in contact with the material. This method will require the firefighter to interact with either the driver or facility personnel to determine where the control valve may be and what the safest method to accomplish the task may be. Do not open or close any valve without understanding its implications.

Emergency shutoffs—transportation. There are three types of emergency shutoffs valves: mechanical, hydraulic, and pneumatic. Closing a mechanical valve closes all internal valves within 30 seconds of activation. Mechanical valves are constructed with corrosion-resistant cables and handles. Closing a hydraulic or pneumatic valve closes all internal valves when a loss of hydraulic or pneumatic pressure occurs. This typically operates off the tractor's system via a pigtail. Emergency shutoffs are usually well marked and located in easy-to-find areas. The shutoff valves are typically found behind the driver's side of the cab or near the control valves.

Figures 4–43, 4–44, and 4–45 are examples of shutoff valve locations on common trailers.

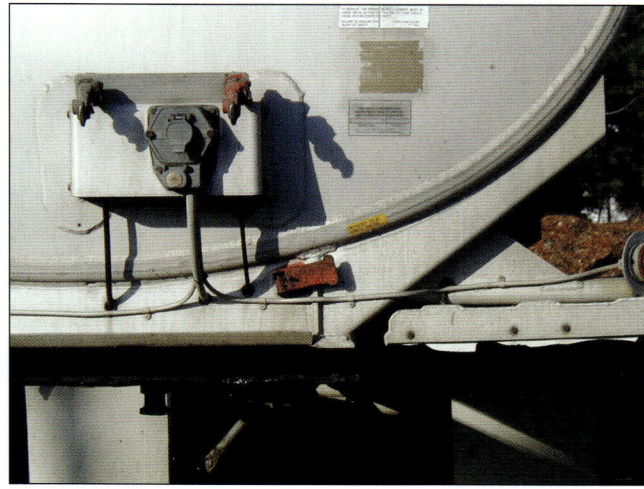

Fig. 4–43. Emergency shutoff in an MC306/DOT406

Fig. 4–44. Emergency shutoff in an MC307/DOT407

Remote valve shutoffs—fixed facilities. Remote shutoff valves in fixed facilities are generally found in facilities with loading and unloading areas (fig. 4–46). They may be located near the entrance to provide easy access to emergency responders. Consult with facility personnel to determine the location. During facility preplans, the responder should look for remote shutoff valves and mark them on a site map.

Fig. 4–45 a and b. Emergency shutoff in an MC331

Fig. 4–46. Remote shutoff valves are generally found in facilities with loading or unloading areas.

DETECTION, MONITORING, AND SAMPLING

HM 6.7 First responders can use a variety of monitors and detectors to determine hazards present at potential hazardous materials incidents. Detection and monitoring equipment range from simple papers that detect presence to sophisticated multi-gas gas detectors that quantify the amount present in the atmosphere. Twenty or thirty years ago only trained hazardous materials teams had access to this equipment but in recent years it has become common practice for fire companies to have radiological monitors, multi-gas detectors and pH paper as part of their standard equipment. It is also common for responders to carry single gas carbon monoxide detectors on medical bags and more and more agencies are conducting post fire atmospheric testing prior to crews removing their SCBAs.

Concentration, dose, and exposure levels

Before attempting to determine action levels it is necessary to understand measurement terms and exposure levels. Common measurement terms are parts per million (ppm), parts per billion (ppb), milligram per cubic meter (mg/m3). Exposure levels include threshold limit value time weighted average (TLV/TWA), threshold limit value short term exposure limit (TLV-STEL), threshold limit value ceiling (TLV-C), permissible exposure level, lethal concentration fifty percent (LC50), lethal dose 50% (LD50), and immediately dangerous to life and health (IDLH).

Parts per million (PPM). Expression used to determine concentration based on a ratio of a solid, liquid or gas in relation to a million parts of a specified medium. For instance, one molecule of chlorine in one million molecules of air or one molecule of acetone in one million molecules of water.

Parts per billion (PPB). Expression used to determine concentration based on a ratio of a solid, liquid or gas in relation to a billion parts of a specified medium. For instance, one molecule of chlorine in one billion molecules of air or one molecule of acetone in one billion molecules of water.

Milligrams per cubic meter (Mg/M³). Measurement for concentration generally used to measure pollutants in the air.

Threshold limit value (TLV). The exposure level of a substance a worker can be exposed to throughout his/her life without suffering any adverse effects. This level is determined by the American Conference of Governmental Industrial Hygienist (ACGIH).

Threshold limit value time weighted average (TLV-TWA). The averaged exposure level of a substance a worker can be exposed to during an eight-hour day without suffering any adverse effects. This level is averaged over an eight-hour workday and forty hours a week. For instance, if the exposure level is 10 ppm and in the first hour you are exposed to 20 ppm and the second hour you are exposed to 0 ppm then your time weighted average is 10 ppm. This level is determined by the American Conference of Governmental Industrial Hygienist (ACGIH).

Threshold limit value short-term exposure level (TLV-STEL). The exposure level of a substance that a worker cannot exceed during a fifteen-minute interval. If this level is reached the worker should be removed for the hazardous atmosphere for at least sixty minutes. Workers cannot meet the TLV-STEL more than four times in an eight-hour shift. This level is determined by the American Conference of Governmental Industrial Hygienist (ACGIH).

Threshold limit value ceiling (TLV-C). The maximum exposure level of a substance a worker can be exposed to during his/her workday. If this level is reached the worker must be removed from the contaminated atmosphere for the rest of the workday. This level is determined by the American Conference of Governmental Industrial Hygienist (ACGIH).

Permissible exposure limit (PEL). The maximum exposure level of a substance a healthy adult can be exposed to during a forty-hour week without suffering any adverse effects. This level was determined by the Occupational Safety and Health Administration (OSHA).

Dose response relationship. The effect of a toxin has on an exposed subject. This effect is dependent on the concentration and duration of exposure. If a subject is exposed to 200ppm for 30 minutes he/she may experience loss of consciousness. However, that same exposure for 2 minutes may have no ill effects but exposure to 2,000ppm for 2 minutes may result in death.

Lethal dose 50 percent (LD50). The amount of a liquid or solid that was fatal to 50 percent of the test animals. Usually the animal group used will be in parenthesis.

Lethal concentration 50 percent (LC50). The amount of a gas that was fatal to 50 percent of the test animals. Usually the animal group used will be in parenthesis.

Immediately dangerous to life and health (IDLH). Exposure level that is likely to cause death or irreversible health effects or would prevent escape from the contaminated environment. This level is determined by the National Institute for Occupational Safety and Health (NIOSH)

Types of monitoring equipment

There is a variety of hazardous materials detection and monitoring devices. This section discusses the function of these devices. Responders use detection and monitoring equipment so they can remain aware of the hazards present at emergency incidents. It is vital to know what equipment is necessary to detect which hazards.

Test strip identification

Corrosives. Corrosives are substances that can attack and cause damage to people, property and the environment. Corrosives are broken down into two categories acids and bases. Acids are substances that release a hydrogen ion when in solution and bases release hydroxide ion when in solution. The amount of hydrogen ions and/or hydroxide ions that can be released determines the strength of the acid of base. The strength of an acid or base is measured by the pH scale using pH paper (figs. 4–47a and 4–47b). Acids that release a small amount of hydrogen ions will have a pH of 5 or 6 and acids that release more or all hydrogen ions will have a pH of 0 or 1. The same principle applies to the bases and hydroxide ions. The other factor that plays a part in the corrosive risk is the concentration. The concentration is the ratio of acid or base compared to the amount of water it is dissolved in. The more concentrated the solution the more damage it will cause. Unfortunately, pH paper can only measure strength but not the concentration.

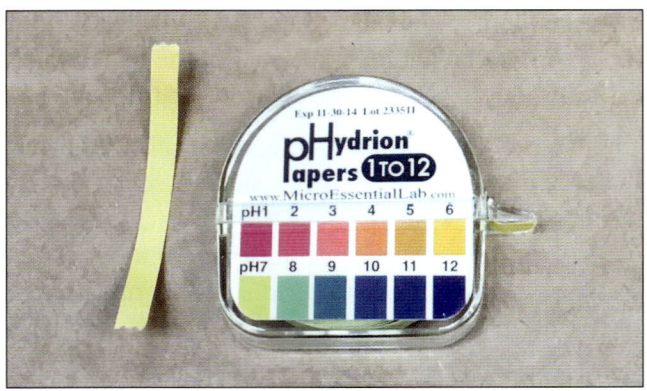

Fig. 4—47a. pH paper

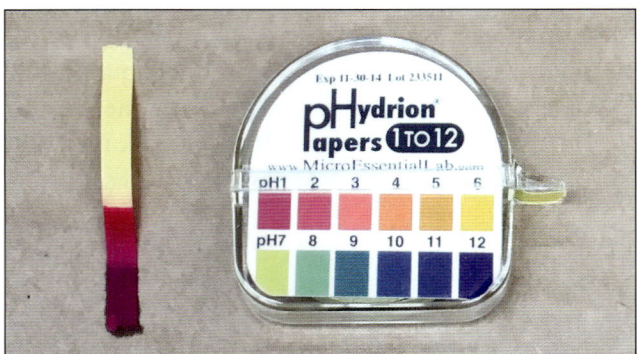

Fig. 4—47b. pH paper detecting an acid

Since corrosives cause damage to human skin and electronic devices it is extremely important to determine a corrosive hazard early in the incident. If you attempt to use electronic meters when a corrosive atmosphere is present the meter is likely to become inoperable. Some corrosives have high vapor pressures and can present a significant atmospheric hazard while others have low vapor pressures and mainly present a contact hazard.

Detecting corrosive hazards is accomplished by using either pH paper or an electronic pH meter. A pH meter is an electronic meter that works by sticking a probe into the material. A pro to using an electronic pH meter is they are very accurate. However, a good unit is fairly expensive in comparison to the paper, you may have to draw a sample. Also, a pH meter needs to be calibrated prior to use and is ineffective for atmospheric testing. Most agencies choose to use pH paper due to its ease of operation and affordability.

The pH scale. pH paper comes in rolls or strips that are chemically treated to change colors in the presence of a corrosive. Corrosivity is measured using the pH scale which ranges from 0 to 14 with 7 being considered chemically neutral (fig. 4–48). Substances with a pH between 0 and 6 are considered acidic with those from 0 to 3 considered dangerously acidic and substances that

range from 8 to 14 are considered basic with those from 10 to 14 considered dangerously basic. While there is not a standard color range for the pH scale we generally see acids range from pink to bright red, neutral is green and bases from green-blue to dark blue. The pH scale slides exponentially, meaning a substance with a pH of 6 is 10 times more acidic than one with a pH of 7, a pH of 5 is 100 times more acidic than a 7 and so on all the way to a pH of 1 is 1,000,000 times that of a 7. The basic side of the scale mimics the acidic so a pH of 13 and a pH of 1 have the same corrosivity. This dispels a myth that acids are more corrosive than bases.

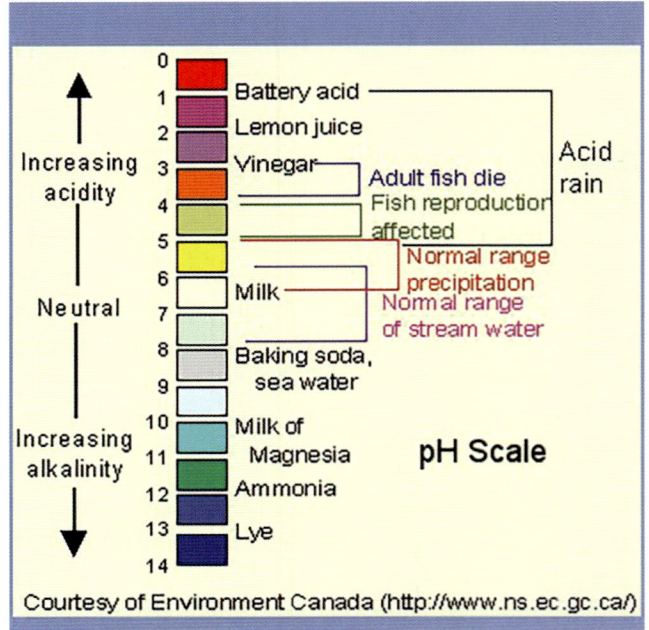

Fig. 4—48. pH Scale

Response objectives. As stated earlier, determining corrosive hazards early is very important. While attempting to monitor ensure that all personnel are wearing the appropriate PPE. To test for atmospheric corrosive hazards, you can tape pH paper to the SCBA facepiece or a stick and observe any color change. If the pH begins to change color responders should report the change to the incident commander and mark the location of the change. This means the substance has a high vapor pressure or the material is reacting with the moisture in the air and will require expanded zones and possible evacuations. Some corrosives are toxic such as hydrofluoric acid and flammable such as Pyroligneous acid so if vapors are present it is prudent to determine any additional hazards and adjust accordingly. If responders are dressed only in structural firefighting gear they should consider evacuation to a safe area. Wetting the pH paper with distilled water will allow it to react to corrosive vapors more quickly than dry pH paper. Many times, responders will use a wetted and a dry piece when approaching the incident. It will be necessary to dip the pH paper in the substance to get the actual pH reading. Most corrosives have very low vapor pressures and will not react to the paper until it is dipped in the substance. This means responders are required to work in close proximity to the substance requiring the use of plash protection with SCBA. Hydrocarbons will sometimes display pH levels of 4 to 6. To get a true pH level on hydrocarbons you will need to wet it and put it in the vapors. If the hydrocarbon has a low vapor pressure wetting the paper will not produce any results.

Other test papers. Besides testing for corrosives there are other chemical test strips that are used to identify other chemicals and WMD's. Test papers are used when encountering unidentified materials or if you are attempting to verify a specific known hazard. All the test strips are chemically treated and will change colors when exposed to the test chemical. As with pH paper responders will be required to dip the test strip in the material which increase the risk of exposure so proper PPE must be worn.

Fluoride test strips.
Fluoride is the simplest anion of fluorine. Fluorine is an element known to react with almost every element in the periodic table including some of the noble gases. It attacks most metals and will cause steel wool to burn. It is a highly toxic, corrosive and reactive yellowish gas. It reacts with many substances including glass, concrete and some rubbers. It is harmful if inhaled and can be readily absorbed through the skin. Hydrofluoric Acid is one of the more common substances made from fluorine. It is highly toxic and corrosive and is absorbed through the skin causing potentially fatal systemic toxicity.

Response objectives. Due to fluoride's highly toxic and reactive nature it is very important to determine if fluoride is present when responders encounter an unidentified material. This is easily accomplished using fluoride test strips where the pink paper turns to yellow in the presence of fluoride. Some substances can cause changes in the paper such as chlorates and bromates. Just like with pH paper, fluoride test strips can be attached to PPE or a stick to detect fluoride vapors. If vapors are not present they the test strip must be dipped into the material. Due to the absorbent nature, it is recommended that responders wear vapor protective suits (EPA Level A).

Water paper.
Water paper simply detects the presence of water in solutions (fig 4-49 a and b). It is a white paper that turns lavender if water is contained. Knowing if water is present helps make certain assumptions about the solution and therefore helps in the decision-making process. For instance, you can assume the solution is not water reactive, is water soluble and a water based decontamination should be effective.

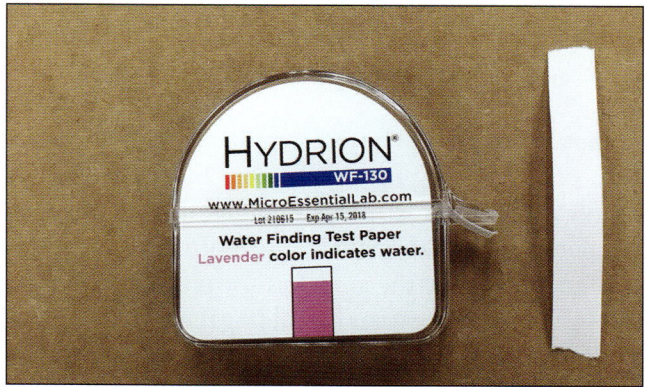

Fig. 4—49a. Water paper

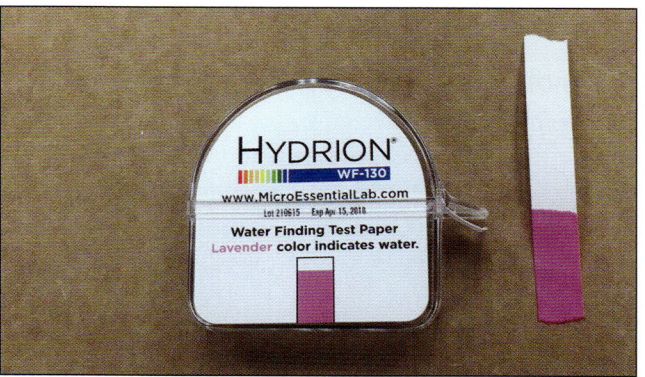

Fig. 4—49b. Detecting water

Oxidizer paper.
Oxidizer paper also known as potassium iodide starch paper will detect the presence of oxidizers and organic peroxides (figs. 50a and 50b). Due to the extremely reactive nature of oxidizers and organic peroxides it is very important to discover the presence as early as possible. Oxidizer paper are white strips that turn to a dark purple or blue if the results are positive.

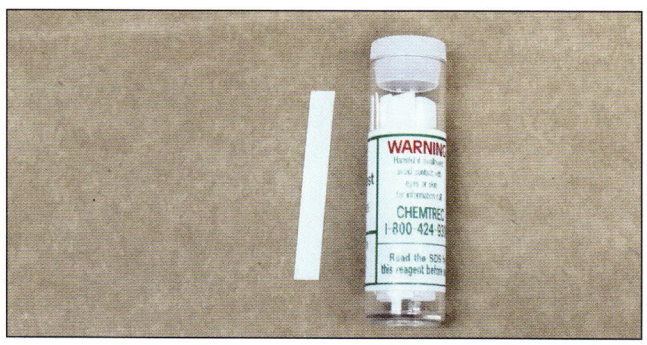

Fig. 4—50a. Oxidizer paper

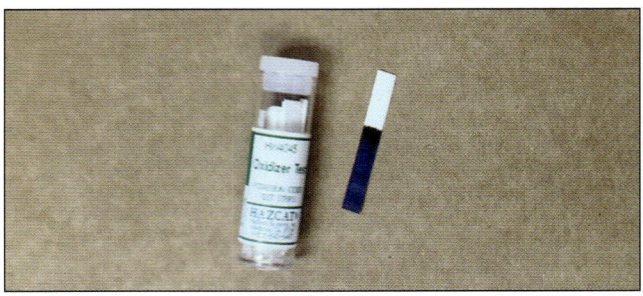

Fig. 4–50b. Detecting oxidizers

Response objectives. Ensure responders are in proper PPE prior to using oxidizer paper. To use on liquids simply dip the test strip into the liquid, for solids you can put one or two drops of hydrochloric acid on the test strip and touch it to the solid. It is also possible to wet the test strip and test for gases by holding the test strip above the liquid. This can also be accomplished by taping a test strip to your PPE or a stick. If the materials tests positive for oxidizers it is imperative to keep the material away from heat or chemical contamination especially combustibles.

Chemical classifier strips. Chemical classifier strips are much like other detection papers where a sample is applied to the detection paper and evaluate the color change. The difference is that chemical classifier strips have multiple test attached to the same strip. For instance, there are the SpylFyter Chemical Classifier Strips that test for various hazards (fig. 4–51). The strips are approximately 12 inches long and one-inch wide. Always check your chemical classifier to ensure it is inside its expiration date. There are five tests on the classifier strip:

- Acid/base (pH)
- Oxidizer
- Fluoride
- Organic solvent/petroleum
- Chlorine/bromine/iodine

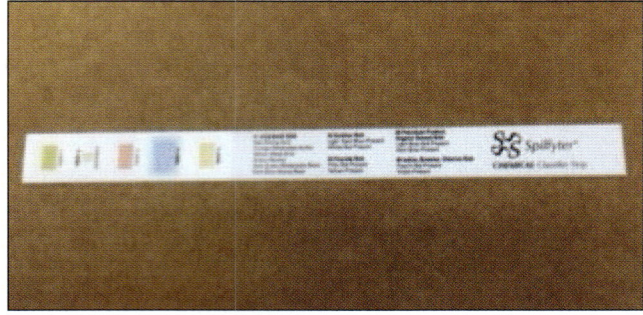

Fig. 4–51. Chemical classifiers

Response objectives. To use a chemical classifier strip, you must ensure you are in appropriate PPE for the situation and approach the spill safely. The reagent papers are in the top half of the strip so always hold the test strip by the bottom. Place the strip face down in the liquid for approximately 30 seconds. If you are testing the contents of a drum tie a string to the hole in the bottom of the strip and lower it into the drum. Once the strip has contacted the liquid leave it there for 30 seconds. The classifier strips come with a chart to evaluate the color change for each test. You can choose to evaluate the strip in the field and leave the contaminated strip in the hot zone for the cleanup contractor. You can also take back into the cold zone and have it read there by the science officer. If you choose to take it back place it in a clean clear container such as a Ziploc baggie so it can be taken back and looked at safely. This will allow you to classify the material or at least rule out certain materials and hazards.

Hazmat Smart Strips are another chemical classifier used to detect airborne contaminates. It is a small rectangle with 8 squares coated with reagents. The smart strip has an adhesive on the back and is meant to be applied to objects or PPE and will activate when exposed to very minute amounts. The smart strip can detect the following:

- Chlorine
- pH
- Fluoride
- Nerve agents
- Oxidizer
- Arsenic
- Hydrogen sulfide
- Cyanide

While chemical test papers can't identify the exact materials, they can classify unidentified materials and provide some answers to what hazards you are facing. They are easy to obtain, affordable easy to use. It is worth evaluating your needs and putting some of these in your detection arsenal.

WMD detection papers and kits

M8 and M9 paper. M8 and M9 papers are used to detect chemical nerve and blister agents (fig. 4–52). While predominantly used by the military we see first responders having this on hand for quick detection if a terrorist incident is suspected. M8 comes in a small book with 25 sheets of paper and M9 comes in a roll with one side being adhesive.

Fig. 4–52. M8 paper

Response objectives. Ensure responder are in proper PPE prior to using M8 or M9 paper. To use M9 a piece is taken off the roll and is placed on objects and personnel that have the likelihood of encountering the chemical agents. If you see a color change using M9 then there is a chemical agent present. M9 can detect presence but is not able identify which agent. To use M8 you tear a sheet out of the book and blot the liquid. Observe any color change and compare it to the chart in the booklet. Yellow – gold indicates a G Nerve agent, dark green indicates a V Nerve agent, Pink – Red indicates a blister agent. If a chemical agent is suspected it is crucial that responders wear appropriate PPE. If you use M9 and have a color change you should use M8 to attempt to identify the chemical agent.

Biological test kits. While there is electronic biohazard monitoring used to detect possible WMD's first responder agencies find it much more affordable to use simple biohazard field test kits. In the early 2000's a number of letters were sent via US mail to reporters and members of congress resulting in 5 deaths and a number of injuries. This raised awareness to the real threat of bioterrorism and the need for first response agencies to enhance preparedness. Since then biological field test kits have been developed to assist first responders make determinations on the presence of proteins on suspicious package and white powder calls (fig. 4–53). The field test kits work by taking a sample of the suspicious powder and using a treated swab determine the presence of biological material. This is identified by a color change. A positive reading will change the test solution to purple. There is also a pH test in the kit as a dual test. Biological agents will usually have a neutral pH so if you receive an acidic or basic test result chances are very unlikely there is a protein present. Also, a positive result for biologic material is not conclusive that it is a dangerous biologic

material. Nonhazardous substances such as cake mix and flour will have a positive test result.

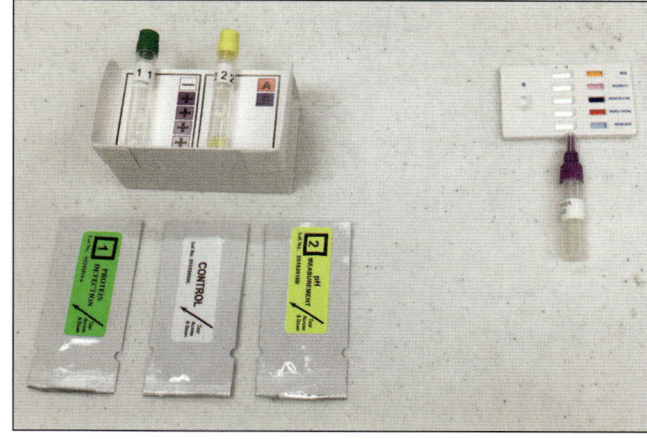

Fig. 4–53. Biological test kit

Response objectives. Ensure responders are in appropriate PPE including SCBA. Notify the local law enforcement agency having jurisdiction. Suspicious packages should be examined by the bomb unit to ensure that no explosives are present. Approach the substance being very careful to disturb the scene as little as possible. If possible take pictures or video of the initial incident scene. Obtain a sample and complete the protein and pH tests per manufacturer's directions and observe results. Responders should leave the field test kit in place and the rest of the scene as you found it and allow law enforcement to investigate regardless of test results. If the protein test is positive then a more agent specific test is necessary and usually conducted at a laboratory.

Colorimetric tubes. Another detection option are colorimetric tubes. There are many different manufacturers for the tube systems and each have their own specific design. Each tube system has tubes with a chemical treated solid substance and a pump used to draw air through the tube (fig. 4–54). There are hundreds of chemical gases, vapors and aerosols that can be detected. Most of the tubes are designed to quantify specific materials although there are some used to classify a group of substances that would be likewise indicated. Each tube has a scale located on the side that has a definitive ppm range (such as 5–100 ppm). You find that jurisdictions have specific hazardous substances that they are unable to detect using any of their detection methods so colorimetric tubes are a perfect fit. The tubes are filled with a chemically treated reagent to produce a color change when exposed to the substance. It is imperative to read the directions for each tube immediately after purchasing and then prior to each use. Each tube has different storage temperatures and use instructions.

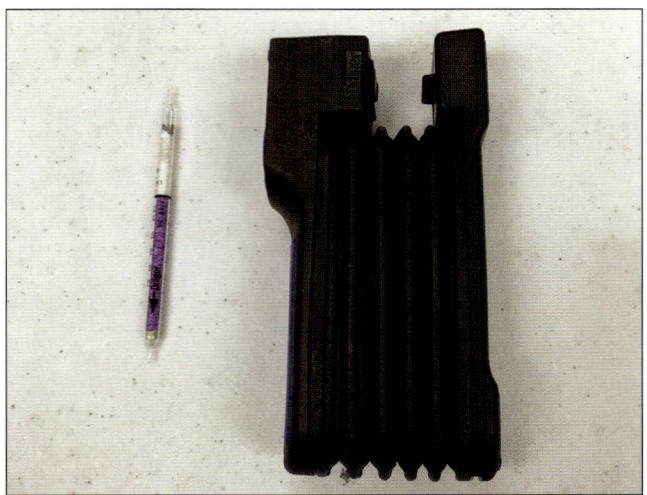

Fig. 4—54. Colorimetric pump & tubes

Response objectives. Ensure responders are in the proper PPE prior to using colorimetric tubes. To use colorimetric tubes, read the directions and prepare the pump and tubes. When you arrive in the area, take out the first tube, the two sealed ends are broken open, then one end of the tube is placed in the pump and the other end allows the air to be pulled through the tube. You deliver the specified number of pump strokes as per the directions. Observe any color change and note the reading using the scale printed on the side of the tube. Note your reading and your location then proceed to the next location and repeat.

Electronic gas detection

Many emergency response agencies already employ the use of either single gas or multi-gas detectors as part of their normal response. For instance, many fire and ems agencies will attach a single gas carbon monoxide detector onto to medical response bags. This will detect the presence of carbon monoxide in structures which can help explain the signs and symptoms of some patients. Or fire departments employ the use of a basic multi-gas detectors for gas leaks, unusual odor calls and post fire air monitoring. These types of units are becoming increasingly valuable to ensure proper response objectives are being selected and for responder safety.

These units are battery operated and comprised of specific sensors meant to determine airborne contaminant levels. Contaminants pass through the sensors and will display concentration levels. Unlike tubes these monitors will continuously monitor and reflect current concentration levels. You will need to understand the response time of your unit and create your monitoring strategy. A response time is the time it takes for the sample gas reading to be displayed. If your response time is ten seconds then you need to move slowly so you aren't in a highly concentrated atmosphere before your sensors can display it.

Single and multi-gas detectors. Single gas detectors are a perfect solution for an agency that has the need to monitor for specific hazards. Fire and EMS may choose to use single gas carbon monoxide detectors for medical calls. Since chlorine is a commonly used gas and can be damaging to a multi-gas detector so having a single gas chlorine detector will fit. Most chemical sensors found in single gas monitors will show concentrations in ppm. Many sensors are available such as chlorine, ammonia, carbon monoxide, hydrogen cyanide. These detectors are small and affordable; however, they do require maintenance. They should be calibrated monthly and a bump check weekly and prior to each use. Calibrations and bump checks should be completed in accordance with manufacturers recommendation. Sensors are rated to last 1–2 years, but life spans vary based on storage and use.

Multi-gas detectors are built and function the same way as single gas sensors but will monitor up to 6 gases (fig. 4–55). Therefore, we will discuss response options for both at the same time. Most fire departments use a standard 4 gas monitor on their fire apparatus but many hazardous materials teams will use a 5 or 6 gas monitor adding a photoionization detector as one of the sensors. The standard 4 gas monitor (confined space configuration) will have the following configuration:

- Oxygen

- Flammability

- Carbon Monoxide

- Hydrogen Sulfide

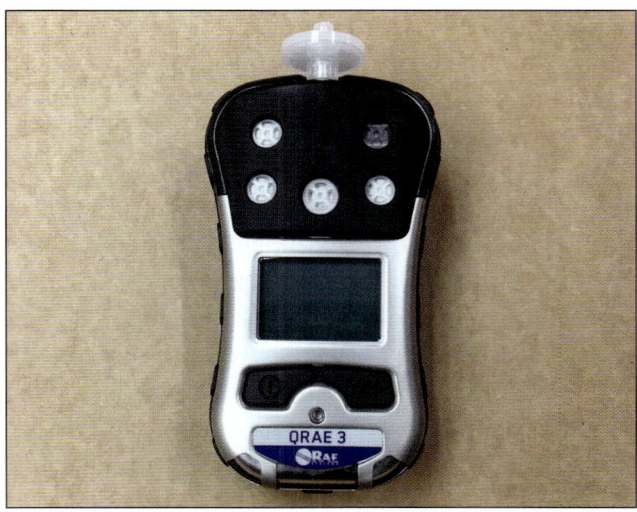

Fig. 4—55. Multi-gas meter

This is considered the confined space configuration due to the hydrogen sulfide which is common in confined spaces. Agencies can select any configuration for their multi-gas monitors but this is the most common and cheapest. Manufacturers sell this configuration in a single cylinder calibration gas. Although many fire departments now switch the hydrogen sulfide for hydrogen cyanide for post fire air monitoring.

Response objectives. Before you deploy a single or multi-gas monitor ensure responders are in proper PPE and you do not have a corrosive atmosphere as this can cause irreparable damage. Try to find the vapor density of the gas so you know whether you should monitor high or low. If the vapor density is less than one then you should hold monitor high if it is higher than one then you should hold the monitor low. If you are dealing with an unknown them you should monitor at all levels. It is also very important to understand how the monitor works what each sensor is telling you. Using a single gas detector such as CO or HCN you should be monitoring in an area where the measurement gas is expected to be or where you are attempting to rule out the presence. The main purpose for using a multi-gas meter in an unidentified hazardous materials situation is to determine O2 levels, flammable atmospheres and VOCs if the unit is equipped with a PID. This is accomplished after it is confirmed that no corrosive atmosphere exists.

Decontamination

Units taken into the hot zone may require decontamination. First rule is to avoid contamination of electronic monitoring equipment. All air monitors will be exposed but may not be contaminated.

Oxygen sensor

The oxygen sensor reads in percentage of air. Normal oxygen level is 20.9% and if the reading drops below 20.9% it means that another substance has displaced the oxygen. If the substance is toxic any drop may be significant. OSHA defines an oxygen deficient atmosphere as that which contains less than 19.5% and an oxygen enriched atmospheres as that which contains more than 23.5% and respiratory protection must be worn in either case. Because of this 19.5% is the low alarm value and 23.5% is the high alarm value. As you will see, respiratory protection may be necessary even at levels outside of the OSHA regulations. If you drop from 20.9% to 19.9% that means you have 50,000 ppm in air concentration of the other substance. We calculate this by understanding some basic math and chemistry. Since we are measuring to the millionth (parts per million), one percent of one million is 10,000. The air is made up of 1 part oxygen and 4 parts nitrogen and since oxygen and nitrogen will displace evenly we must assume that 1 percent of the oxygen and 1 percent of the nitrogen is being displaced. If 10,000 ppm of oxygen is being replaced then four times that amount of nitrogen is being displaced or 40,000 ppm totaling 50,000 ppm. If the material displacing the oxygen is hydrogen cyanide and hydrogen cyanide has an IDLH value of 50 ppm then you are at 1,000 times the IDLH level which is lethal with a short exposure. Also, if you drop from 20.9% to 20.8% oxygen you have a 5,000 ppm in air concentration which is still 100 times IDLH for hydrogen cyanide. So even the slightest variation should be noted and reported to command. Oxygen enriched atmospheres generally mean that an oxidizer or organic peroxide is present. In oxygen enriched atmospheres, there is a significant risk of ignition and fire. Materials that may not burn under normal circumstances may burn vigorously in an oxygen-rich environment and materials that burn normally may burn much hotter and faster. Fires involving oxidizers usually produce extremely toxic by-products. For instance, a methane fire produces carbon dioxide and water but if you add chlorine to the mix you will hydrogen chloride and carbon tetrachloride. Special precautions should be taken when encountering this hazard.

Combustible gas sensor

Combustibility sensors detect the presence of flammable gases and display in a percentage of the lower explosive limit (LEL) of the calibration gas. Most combustibility sensors are calibrated with methane or pentane. The LEL of methane is approximately 5% in air concentration. If your sensor is calibrated to methane and it reaches 100 then you are at the LEL or you have a 5% in air concentration present. The low alarm for the combustible gas sensor is 10% of the LEL and the high alarm is 20% of the LEL. However, because combustible sensors are only accurate when measuring the calibrations gas then you will need to find the correction factor to determine a correct reading. You should get a correction factor chart form the manufacturer. If you are measuring propane and are calibrated to methane the correction factor is 1.6. To receive an accurate meter reading simply multiply your actual reading by 1.6. If your meter is reading 10 or 10% of the LEL and your correction factor is 1.6 then your actual reading is 16%. Unfortunately, there are some inherent issues with combustible gas sensors.

These sensors need proper oxygen levels to function properly. If you find yourself in oxygen deficient or enriched atmospheres you may get inaccurate results. If the flammable gas concentration is too high you may saturate the sensor which may require you to calibrate before continuing use. Also, saturating your sensor reduces the overall life of the sensor. Exposure to certain chemicals will cause he sensor to become poisoned which will result in at least requiring a recovery period in clean air before the sensor is functional to causing the sensor to completely fail. Substances that can poison these sensors are halogen and materials that contain silicone.

Photoionization detectors

A photoionization detector (PID) measures volatile organic compounds (VOCs) and some other toxic substances and display them in either parts per billion (PPB) or parts per million (ppm). PIDs work by using an ultraviolet light to change molecules into positively charged ions. Molecules pass through the ionization chamber and absorb the energy of the UV rays which will displace negatively charged electron from their orbit and create a positively charged ion. The ions are measured by the detector and displayed. There are various sized UV lamps available for PIDs, 9.8ev, 10.6ev and 11.7ev. Most agencies use the 10.6ev bulb in their units because the 9.8ev can only detect a very small amount of substances and the 11.7ev bulb is very expensive and has a lifespan of only 2-6 months. All substances ionize at different levels so your PID will only be able to detect substances with an ionization potential (IP) of less than your bulb size. For instance, the IP of ammonia is 10.18ev, if your PID has a 10.6ev bulb your PID will be able to detect it. However, the IP for chorine is 11.48ev so an 11.7ev bulb will be required to read it. Much like the combustible gas sensor, a PID is only accurate when measuring its calibration gas. Most PIDs are calibrated with isobutylene and require correction factors to obtain accurate readings with substances other than isobutylene. Therefore, you will need to obtain a correction factor chart from the manufacturer. Agencies can choose to get a standalone PID or have the sensor placed in a multi-gas meter (fig. 4–56).

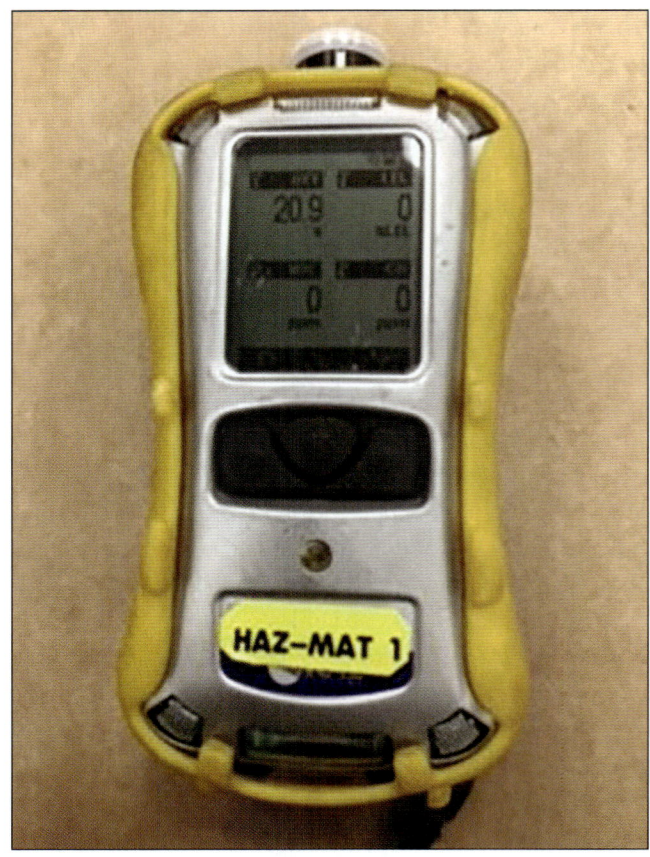

Fig. 4–56. Multi-gas meter with PID

Radiological monitors

Radiation has many uses, it is widely used in the medical community, energy production and naval propulsion. Unfortunately, radiation is also one of the most feared weapons and excessive exposure to radiation can cause great harm. There is a fear that terrorists will use a nuclear device in a terror attack but it is much more likely to use radioactive material and a conventional explosive to disperse the material over a large area causing mass contamination, this is called a radiological dispersion device (RDD). Due to its many uses radioactive material is commonly transported by rail and over the road. It is very important that first responders understand the hazards of radiation, ways to detect it and plans to protect themselves and the community from accidental or criminal exposure. Below are the different kinds of radiation detection instruments available to agencies.

- **Dosimeters** display the accumulated dose of responders.

- **Survey meters** detect the presence and dose rate but not the type of radiation (alpha, beta, gamma or neutron). Survey meters come with different probes and each has its uses and limitations. Some probes are used to detect and measure dose rates and others are used to detect contamination.

- **Isotope identifiers** will detect the dose rate and which isotope you are encountering.

Most agencies have basic dosimeters and survey meters (fig. 4–57). Isotope identifiers are not as common in the response community due to the cost and survey meters have been found to be effective for most of the response community.

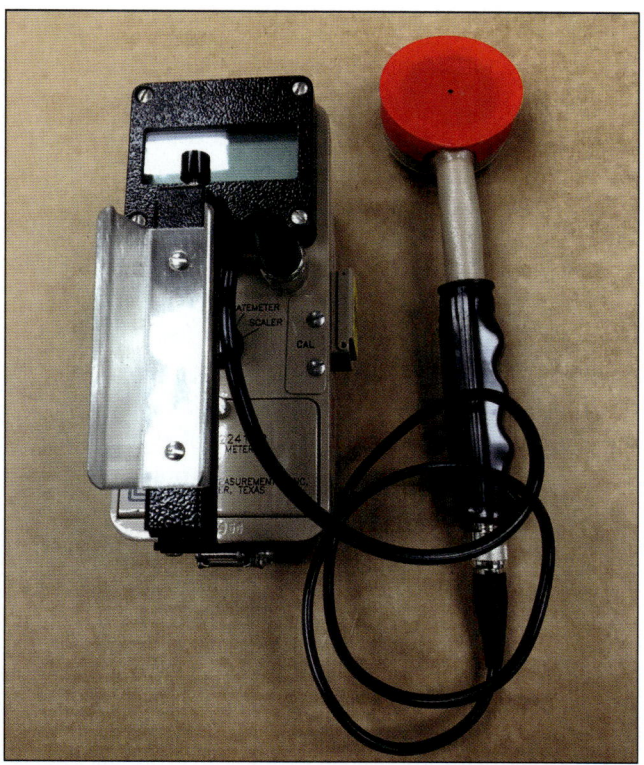

Fig. 4–57. Survey meter and dosimeter

Units of measurement. Radiation has a couple units of measurement depending on the circumstances and what you are looking at. Below are the common units of measurement in the United States:

- **Roentgen:** a unit of exposure to radiation. Roentgens only apply to gamma and x-ray radiation and based on the amount of exposure will be displayed in either roentgen or R, milliroentgen (one thousandth of a roentgen) mR, or microroentgen (one millionth of a roentgen) μR.

  - Accumulated dose: the amount of radiation an object or person has been exposed to.

  - Dose rate: the amount of radiation an object or person is subjected to per unit time. For instance, if you're exposed to a dose rate of 100

mR/hr for 15 minutes then your Accumulated dose is 25 mR.

- **Absorbed dose:** displayed as rad and is the amount of radiation energy absorbed by the medium or object that it has passed through. The absorbed dose does not differentiate between types of radiation (alpha, beta, gamma, or neutron).

- **Equivalent dose:** displayed as rem and is the effect of an absorbed dose to those exposed.

Radiological monitoring technology. First there is the Geiger Mueller Tube technology (GM). Monitors using GM technology have tubes filled with gas. The radiation passes through the tube and using high voltage electrical currents ionizes the gas inside the tube and the detector measures the amount of ions created. This type of detector can measure different types of radiation (alpha, beta, and gamma). This is the most common type of detector technology used in emergency services. Also called a Geiger counter and has been around since the late 1920's. Then there is the Scintillation Crystal technology. The detector has a crystal, such as sodium iodide or zinc sulfide crystal. When radiation hits the crystal, it creates a light that travels through a multiplier that intensifies the light energy and dislodges several electrons. This is measured to determine the amount of radioactivity. Generally, scintillation crystal detectors can only measure one type of radiation based on the kind of crystal. For instance, sodium iodide crystals measure gamma and zinc sulfide crystals measure alpha radiation.

Response objectives. Ensure responders are in proper PPE prior to using radiological monitors. In addition to incidents that have a radioactive marking or placard it is recommended that emergency responders monitor for radiation anytime they respond to suspected terrorism and any incident involving explosives. In the event of a transportation incident responders should focus on notification of the proper authorities based on your state's radiological incident policies, isolation of the incident, removal and decontamination of any exposed people and protection of responders and the public. Emergency worker exposure guidelines have been established and are as follows:

- 5 rem maximum for all occupational exposures

- 10 rem maximum for extraordinary circumstances or protecting critical infrastructure

- 25 rem for life saving activities

- Exposures above 5 rem are on a voluntary basis

Hazardous Materials Operations: Mission Specific

HAZMAT

Chapter 4

Ensure responders are in the appropriate PPE including SCBA. The best protection methods for radiation exposure are time, distance and shielding. Limiting time in the contaminated area, maximizing distance from the source and utilize proper shielding depending on the type of radiation (paper for alpha, plexiglass for beta and lead for gamma).

Assessing needs and evaluating results

Using information to assess monitoring needs.

Monitoring and detection is a very crucial part of the hazardous materials incident. Once responders arrive at the incident they begin to gather information about the product and the container. Using clues such as placards/labels, shipping papers and visual clues they begin to make initial decisions such as isolation perimeters and begin gathering information like chemical and physical properties. From here responders can begin predicting hazards and product behavior. Estimations on zone size and public protection option. However, all decisions made up to this point are merely speculative because none of the information has been confirmed. It is now responders can deploy detection and monitoring devices (fig. 4–58). All monitoring and detection activities must be accomplished in the appropriate PPE to include respiratory protection.

Fig. 4–58. Air monitoring

Determining monitoring strategy is based on the individual incident and materials involved. For instance, if you have an overturned MC412 with a placard of 1830 (shipping papers confirm sulfuric acid 98%), has a vapor pressure of 0.001 nnhg, vapor density of 3.4 and an expected pH of 0-1. This product is not flammable, will burn the skin and eyes and is primarily toxic by ingestion and inhalation of mist. The tanker is leaking and the liquid is pooling approximately 40 ft from the tanker. Based on general information it is a corrosive so using the multi-gas meter will not only fail to yield results it is potentially damaging. The detection type best suited for this material would be pH. Based on the vapor pressure we won't expect much in the way vapor development but what vapors are produced will be found low to the ground based on the vapor density. Responders don PPE and conduct atmospheric pH tests then a dip test to confirm the pH. Due to the lack of vapor production the hot zone will not extend beyond the spill area and personnel should avoid contact.

Another example would be a derailment with a railcar containing acrylonitrile, inhibited. The railcar is actively leaking. This product has a vapor pressure of 84 mmhg at 68°F (20°C) and since it is 98°F (37°C) today we know it would be significantly higher. Therefore, the prediction of product behavior would be a significant amount of vapor production. The vapor density of this product is 1.9 so the vapors will stay closer to the ground and spread out. It has a flashpoint of 30°F (–1°C), an LEL of 3% and UEL of 17% so it is very flammable. It has an IDLH of 85 ppm and it's toxic by inhalation, ingestion, contact and skin absorption. We can expect the pH to be neutral but its ionization potential is 10.91 and our PID bulb is 10.6 ev. Therefore, our PID will be unable to detect it. Unlike the last incident which was very straight forward, the Incident Commander has many decisions to make. This product is highly toxic, flammable, will produce a significant vapor cloud that stays low and spreads out and people are susceptible via all four routes of entry. As for monitoring and detection, a multi-gas detector is appropriate but will only hit on the LEL. However, based on calculations it can be estimated that the multi-gas meter will not detect levels below 300 ppm which is almost 4 times IDLH. Our only other monitoring option would be a single gas meter or colorimetric tubes if they are available.

In the event of an unidentified material such as an abandoned drum, responders will need to use all monitoring and detection available to them. This gives them the best chance of ascertaining useful information. It would be wise to start with either pH paper or a chemical classifier to see if you get a hit on vapors. The next step would be to dip test the chemical classifier and document the results. Once that is accomplished and corrosive vapors are ruled out then deploy the multi-gas meter and PID. If the material causes the LEL or PID to display a reading but the identity of the material is unknown then you can only detect the presence but the reading on the meter will not be accurate since the correction factor is not known.

Sampling

Ensure responders are in the proper PPE prior to conducting sampling activities. Sampling for gases has already been covered when atmospheric testing was discussed. One of the hazards of sampling is the need to come in contact with the product. In order to conduct monitoring and detection it may be necessary to obtain a solid or liquid sample. Samples should be placed in either plastic or glass containers. Sometimes samples need to be taken out of the hot zone for further testing. Liquids samples can be taken by using a pipette and a jar or bottle (figs. 4–59a and 4–59b). Solid samples can be taken by using a scoop or shovel and placing the sample in a jar or bottle. The sample jar must be decontaminated prior to exiting the warm zone. Sampling equipment can either be left for the cleanup contractor or decontaminated and returned to inventory.

Fig. 4–59b. Drum sampling

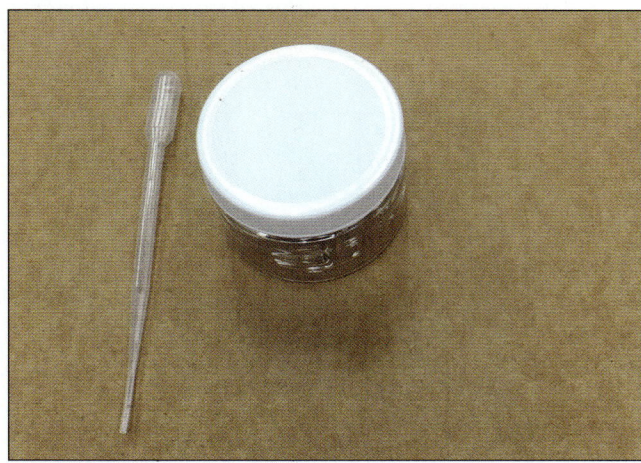

Fig. 4–59a. Jars and pipette

VICTIM RESCUE AND RECOVERY

Objectives

HM 6.8 Upon completion of this section, you should be able to do the following:

- Describe the setup of victim rescue operations and recovery operations. (NFPA 1072, 6.8.1)

- Explain the methods of victim rescue. (NFPA 1072, 6.8.1)

- Explain the types of monitoring equipment. (NFPA 1072, 6.8.1)

- Describe documentation and reporting requirements for victim rescue and recovery. (NFPA 1072, 6.8.1)

Victim rescue is the saving of those exposed or injured during an incident while victim recovery involves the transfer of deceased victims from the hazard zone to the morgue. If the incident is a suspected crime scene it may be necessary for deceased victims to remain as found until the scene is processed. Since life safety is the incident priority victim rescue is the first consideration. In contrast, victim recovery is considered only after hazard mitigation and is usually accomplished well into the incident at the direction of law enforcement and the coroner. Rescue operations can range from a single exposed or injured victim to a mass casualty and the incidents can range from accidental releases from vehicles or industry to terrorist acts. Operations level personnel trained to the mission-specific operation of victim rescue and recovery must operate under the guidance of a hazardous materials technician, an allied professional an emergency response plan or standard operating guideline.

Determining victim viability and conducting triage

Victim viability involves a few main factors:

- Victim condition (ambulatory vs non-ambulatory)

- In line of sight or not in line of sight

- Severity of injury or exposure

If the victim(s) are ambulatory and in the line of sight responders should attempt to communicate with via an apparatus PA or bullhorn and direct them away from the hazard zone to a safe area at the edge of the hot zone to wait for decontamination, medical treatment and transport. Victims who are in line of site but non-ambulatory create more of a challenge for responders. A complete hazard analysis should be completed to determine if the available PPE is adequate. Part of the decision-making process is whether the victims are in the open and can be easily extricated of if they will require any type of extrication or special rescue equipment. Once a rescue team has been assembled they will make contact with the victims and try to determine if there are signs of life. Since it is very difficult to conduct patient assessment while in PPE the rescue team should attempt to elicit a response. Victims showing signs of life should be packaged and transferred to decontamination and passed on to the medical team.

The next group of victims would be the ambulatory victims who are not in line of sight. This is going to require a rescue team to attempt to locate this group of victims. Once located, rescue team members can direct these victims to the safe area to await decontamination, medical treatment and transport. The non-ambulatory victims who are not in the line of sight are the least viable

and difficult to rescue. Rescue team(s) must conduct a search and once located attempt to determine any signs of life. These victims have usually been subjected to significant amounts of contaminant and have the lowest chances of survival.

Determining victim priority based on the severity of their medical condition is called triage. This system is usually conducted during mass casualty incidents. This is a tag system that prioritizes patients from the first treated to the last treated (fig. 4–60). It is a tag system that uses the following ratings:

- **Red tag.** Highest priority and signifies critically injured or ill, treated first

- **Yellow tag.** Second highest priority and signifies seriously injured or ill, treated next

- **Green tag.** Third highest priority and signifies minor injury or ill, treated next

- **Black tag.** Last priority and signifies deceased victims

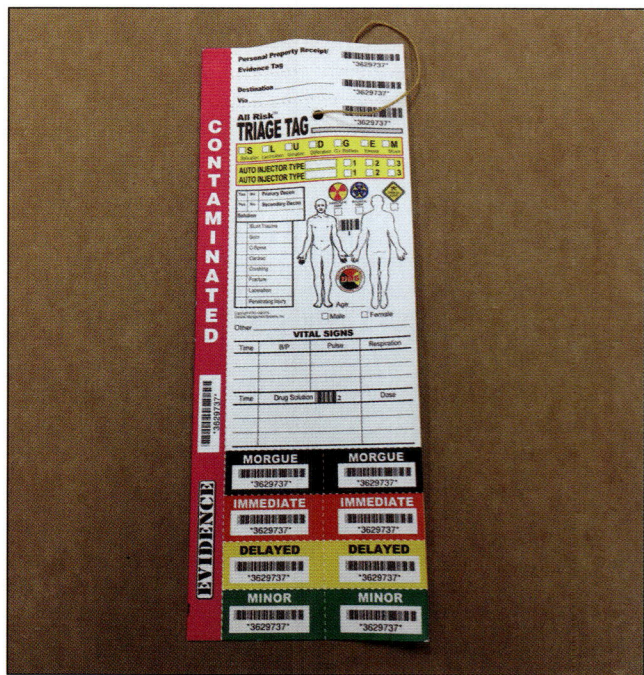

Fig. 4–60. Triage tags

Determining the feasibility of rescue

The feasibility of conducting a victim rescue at a hazardous materials incident is complicated. Emergency services personnel are trained to get off the apparatus and jump into action. However, hazardous materials incidents are more complicated and require a more controlled response. Early in the world of hazmat everyone was taught the "rule of thumb". You extend your arm with your thumb extended in a thumbs-up fashion pointed toward the incident, close one eye and if the scene is visible at all around your thumb you are too close. As responders learned more we have found that while dangerous, the degree of hazard is based on the physical and chemical properties of the contaminant. Therefore, the command staff must make a hazard analysis before making decisions at incidents.

To conduct a hazard assessment an attempt to identify the contaminant through markings, placards and shipping papers, if available. Also, if possible, a site survey using air monitoring equipment, reagent papers, colorimetric tubes is recommended. Once an initial hazard analysis is completed command staff can decide if the responders currently on scene have the appropriate numbers, equipment and PPE to conduct victim rescues. This can be accomplished in a very short period of time. Also, the condition of the victim plays in the decision as does the location of the victim. Many times, the first responders on scene initially are limited to structural firefighting gear and SCBA as their PPE. Since this level of PPE is not made for chemical resistance using these personnel as a rescue team is a difficult decision for command staff to make. After the arrival of the hazardous materials team arrives and is established proper rescue teams can be assembled and proper PPE is available but it may be too late.

Many times a rescue of line of sight non-ambulatory victims is attempted by the first engine company. They will operate in a snatch and grab type rescue operation on victims who are in the open. While structural firefighting gear is not designed for chemical environments it does provide limited protection. These are decisions made by experienced personnel using all information available and instructions to the rescue team to stay out of the product. Agencies should have policies and procedures that apply to victim rescue. A viability and feasibility process should be determined based on equipment and staffing. Also, rescue training should be conducted on a regular basis. This will provide company officers with some guidance by which necessary quick decisions can be made.

Rescue teams and necessary equipment

Rescue team(s) are assembled and based on the incident conditions could require a variety of positions. You will have a rescue team leader which has the following responsibilities:

- Team assembly
- Ensuring proper PPE
- Communicating the team objectives
- Assembling the necessary equipment

You will also have a victim packaging team who will bring the packaging equipment and package and transport viable victims to decontamination (figs. 4–61a and 4–61b). Sometimes a site assessor will be assigned to use monitoring equipment and characterize the site.

Fig. 4–61a. Victim packaging

Fig. 4–61b. Victim removal

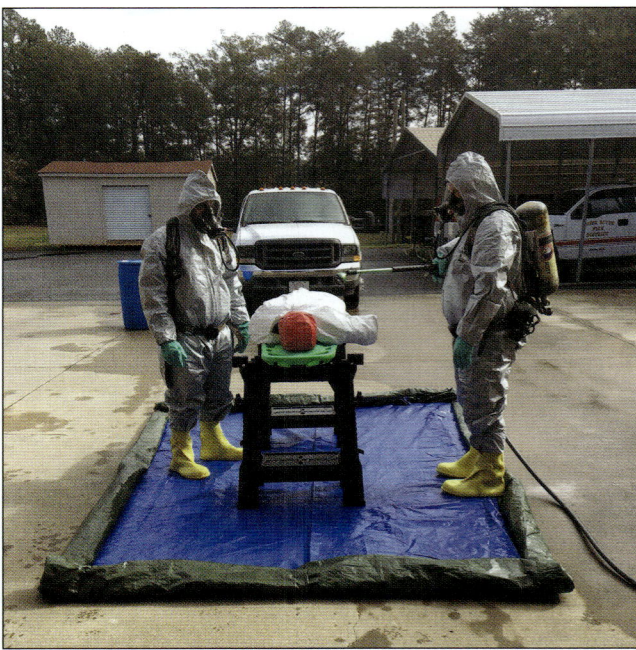

Fig. 4–62. Emergency decontamination

Rescues involving non-ambulatory victims or responders are more complicated and will require special equipment. Also, a formal emergency decontamination line is advisable if the personnel to staff it are available (fig. 4–62). To conduct formal rescues may require the following equipment (fig. 4–63):

- Air monitoring devices, papers and tubes

- Sked® or stokes baskets

- Golf cart, gator or pull carts

- Roller systems

- Pneumatic or hydraulic tools

- Saws

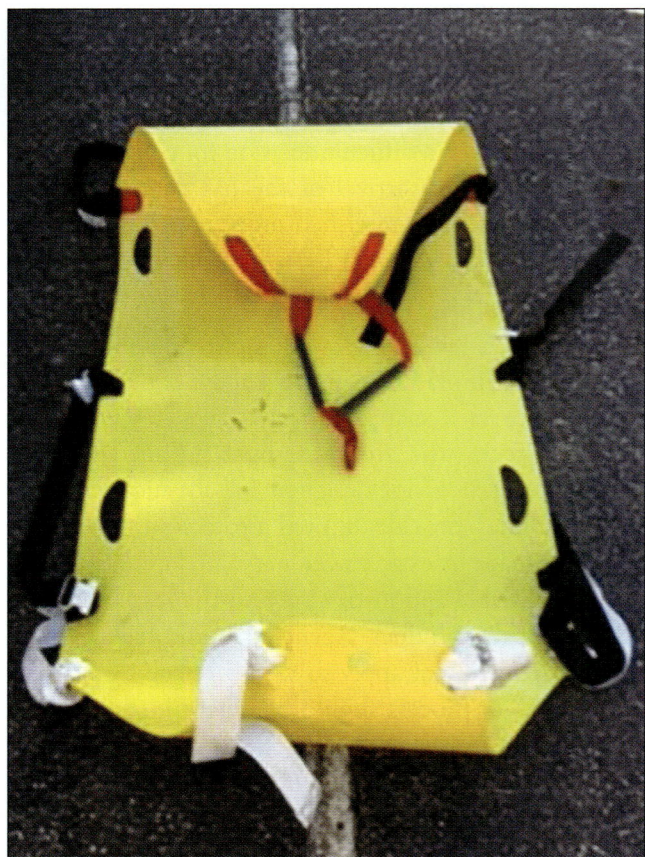

Fig. 4–63. Rescue equipment

Victim recovery

Victim recovery is the removal of deceased victims. This stage will come nearer to the end of the incident, after the hazard has been mitigated. The coroner and law enforcement will be involved. There will most likely by a police investigation and it is recommended that all victims should not be disturbed until after the investigation. If a body has to be moved then photographs of the incident should be taken and turned over to local law enforcement.

Documentation

The complexity of the incident will determine the amount of documentation required. Each AHJ will have different reporting and documentation requirements. Smaller incidents may only require NFIRS reports, also some agencies will want patient forms completed. Larger incidents may require NIMS reports in addition to the NFIRS. All hazardous materials incidents require site safety plans and many agencies require a formal incident action plan.

RESPONSE TO ILLICIT LABORATORIES

HM 6.9 Illicit labs have become a growing problem across the United States in recent years. When someone mentions illicit labs thoughts immediately go to drug manufacturing, but there are many other types of illicit labs. Examples of illicit labs can include:

- Drug manufacturing
- Explosive manufacturing
- Chemical warfare manufacturing
- Radiological device
- Biological specimen growth

In addition to standard hazards present at hazardous materials incidents each of these laboratories present unique hazards and they all must be treated as crimes scenes. All personnel operating at the incident will be required to wear appropriate PPE based on the hazards identified such as flammability, corrosivity, oxidizers, chemical, biological or radiation. Appropriate PPE will be selected and approved by the command staff. This will range from structural firefighting gear to chemical protective clothing. Respiratory protection can range from APRs to SCBAs. Also, there can be additional hazards at illicit labs such as booby traps designed to target intruders that includes first responders and some will have armed persons. This will often require law enforcement to wear body armor and members of the bomb squad to wear blast suits. These additional hazards must be eliminated prior to hazardous materials response personnel entering the hot zone.

At illicit labs, hazardous materials response personnel need to operate in conjunction with law enforcement to mitigate the hazards and process the crime scene. Some jurisdictions train law enforcement personnel to don and work in CPE so they can better process the scene while others depend on hazardous materials responders to mitigate the hazards, provide for decontamination and to collect/preserve evidence. For this type of cooperation hazardous materials responders and local law enforcement should conduct joint training exercises. Operations level personnel trained to the mission specific operation of Response to Illicit Laboratories must operate under the guidance of a hazardous materials technician, an allied professional including law enforcement or others with similar authority, an emergency response plan or standard operating guideline.

Recognizing illicit labs

Not all illicit labs are reported as such to 9-1-1 or provided by informants. Many times, illicit labs are involved in fires or those operating the lab are overcome with chemicals used and are found unconscious. Therefore, all responders should be familiar with common indicators for illicit labs.

If responders find packages of ephedrine or pseudo-ephedrine, acetone, red phosphorus, lithium batteries, or lab equipment such as beakers, flasks glass tubing, burners or compressed gas cylinders where the valves have turned green there is a good chance they have happened upon a drug lab. Due to the numerous methods to make narcotics the list of possible chemicals is extensive. Local law enforcement should be consulted to provide training on the types of labs commonly found in any specific area. The chemicals used in drug labs are, for the most part household chemicals and therefore are easier to protect against.

Another common illicit lab is the explosive manufacturing lab. The manufacturing of explosives can occur in a garage or shed and many of the ingredients are easy to come by such as gun powder, fireworks, pipes and fuses. People sometime manufacture pipe bombs to remove stumps from their property or for entertainment. This is still illegal but not all explosives found are linked to terrorists. To make larger and more sophisticated explosives you may find oxidizers or organic peroxides such as ammonium nitrate and benzoyl peroxide. You may also find jars of urine to make potassium nitrate. Other items will also be necessary such as blasting caps and electronic switches.

The remaining types of labs are much less common but no less dangerous. They require particularly dangerous chemicals and agents which makes it harder for responders to protect themselves. Finding how to manuals is easier than ever today, however, getting your hands on some of the materials can prove challenging.

Biological labs you may find:

- Petri dishes

- Incubators

- Castor beans

- Proteins

- Tissue
  At chemical warfare labs you can find:

- Sodium cyanide, cyanogen chloride and hydrogen cyanide used to make blood agents

- Organophosphorus and rubbing alcohol is used in the manufacturing of nerve agents

- Sophisticated lab equipment

Biological and chemical warfare agents have the potential for massive kill rates however, due to the extreme dangers it poses for the manufacturers and the need of a sophisticated lab set up not many terrorist organizations are successful in the production of these agents.

Radiological labs are extremely rare and are not commonly used to manufacture weaponized radiological material. For the most part, that danger is reserved for governments. The purpose of these labs is to take radiological material and create radiological dispersal devices. These are explosive devices that contain radioactive material. The explosive force will disperse and contaminate the surrounding area with radiation. In radiological labs you could find:

- Radiological detection equipment

- An abundance of smoke detectors

- Industrial sources

- Common materials that are radioactive such as minerals and some pottery

The best sources are radioactive fuel or weapons grade sources but those are extremely difficult to obtain.

Once fire and/or EMS personnel suspect the presence of an illicit lab they should immediately notify law enforcement. Also, due to the possible chemical, biological or explosive hazards that may be present they should retreat to a safe area until a proper hazard analysis can be completed. The only exception to this could be if they are conducting a rescue. Rescuers should be in the greatest level of PPE available and the action approved by the incident commander. Once the rescue is complete the site should be secured and personnel evacuated. Common hazards at illicit labs include flammability, corrosivity, toxicity, radioactivity and infectious materials.

Depending on the nature of the incident will determine who is notified for investigative purposes. Initially, local law enforcement should be notified and they will be responsible for crime scene management. Crime scene management can include rendering the incident safe, scene processing, evidence collection and preservation as well as notification of state and federal agencies. Many different agencies are authorized to investigate criminal incidents. Some illicit labs are handled by local law enforcement, while others, such as large drug labs may include state law enforcement or the Drug Enforcement Agency (DEA). Larger labs can include environmental crimes and the investigatory agencies should include the state health department and the Environmental Protection Agency (EPA), acts of terrorism are investigated by the Federal Bureau of Investigation (FBI). Therefore, if a chemical warfare of biological agent lab is suspected local law enforcement should contact the FBI immediately. For explosive incidents local law enforcement may bring in the Bureau of Alcohol Tobacco and Firearms (ATF) and or the FBI. Once the incident has been declared a crime scene, all state and federal notifications should be done by local law enforcement.

Response objectives for crime scenes

Once an illicit lab has been discovered and has been declared a crime scene it adds a new and exciting dynamic to it. Due to the increased complexity, incident command switches to a unified command post. Unified command is a system where a representative from each agency is present in the command post and command responsibility is shared depending on which role is in the lead.

Initially law enforcement takes the lead to render the scene safe from hostile attack, booby traps or explosives.

Once that is accomplished, life safety and hazard mitigation should take the lead. This includes rescue, treatment and transport of injured victims, evacuation/shelter in place, mitigating the hazardous materials and providing decontamination. The first part of hazard mitigation is locating material(s) and hazardous materials responders will attempt to identify hazards at the site. This is not an identification of the material but a hazard classification of the material such as radioactive, flammable or corrosive. Hazard classification is usually completed on site through monitoring strategies. During this phase, hazardous materials responders are in appropriate PPE and use radiation detectors, air monitoring equipment and chemical/biological detection papers or tubes. Hazardous materials workers should take extra care to keep the scene as undisturbed as possible or take photographs throughout the process. With the hazards identified, hazardous materials responders can provide for victims, ensure decontamination is adequate and the site can be processed in a relatively safe environment.

When all victims are taken care of and the hazard is mitigated crime scene processing begins. Hazard mitigation means that it is no longer a threat to the public. It doesn't mean that chemical protective clothing is no longer necessary. That doesn't happen until has been cleaned up. This is the time when the decision is made to have law enforcement personnel process the evidence and crime scene in CPC or have hazardous materials personnel collect and preserve evidence.

Either way, law enforcement is responsible that all necessary search warrants are obtained. Also, if the evidence gets contaminated or a proper chain of custody isn't kept, valuable evidence may be found inadmissible in court so it is imperative the scene is processed correctly and proper sampling procedures are followed. This begins with scene security. Besides the normal hot, war and cold zones, law enforcement must establish a crime scene perimeter and document everyone that has entered the crime scene.

Collecting and preserving evidence

The first thing that has be understood is that all evidence must be accounted for at all times. This is accomplished by tracking evidence from the time it is collected until it is secured in law enforcement evidence storage. This practice is known as chain of custody. Each agency has a chain of custody policy and it must be followed. When the evidence is discovered and packaged a chain of custody for is filled out. That form must follow the evidence until it is disposed of or it is returned. The chain

of custody form records the name of the person taking custody along with the date and time of transfer, sample ID#, description, and the evidence location. If there is a gap in chain of custody that evidence may no longer be viable for court.

Crime scene processing is conducted through a scientific method where investigators will develop a hypothesis and work the scene trying to prove it. Evidence identification should be accomplished by qualified investigators because to the untrained eye many pieces of evidence will be overlooked. During hazard mitigation, a hazardous materials responder may find it necessary to collect or move potential evidence in order to preserve it. If this occurs, it is crucial that any actions taken be meticulously documented. All evidence samples obtained must follow local, state and/or federal collection, packaging and transportation guidelines. All air monitoring readings must be recorded and all detection papers should be labeled, catalogued and sealed as evidence. All sampling equipment must be clean prior to use. This means that if you have multiple materials or are taking samples from multiple locations a fresh set of equipment must be used for each. For collecting evidence, special sampling equipment and containers may be required. These items are sealed and certified as clean. The lead law enforcement agencies policies will dictate the types of equipment and containers necessary for obtaining samples.

Decontamination

Determining proper decontamination is the responsibility of the command staff. For the most part, decontamination methods at illicit labs are much like the methods at most hazardous materials incidents. As with most aspects of illicit labs there are a few unique matters that must be considered.

The sampling team must take all sealed and labeled samples through decontamination and have the exterior of the sample containers decontaminated. To maintain chain of custody the sampling team should remain in possession of the containers throughout the decontamination process or at least in sight of the containers.

After decontamination, the sampling team must deliver all samples to the law enforcement personnel responsible for evidence collection. Labels are verified and chain of custody forms are completed before the sampling team is permitted to part with the samples. Transportation to the receiving lab is the responsibility of the law enforcement agency having investigative authority. Keep in mind, local, state and federal law enforcement has different sampling and labeling policies.

Personnel involved in any evidence collection or preservation is subject to testify in court.

There are also special decontamination considerations at crime scenes. Decontaminating suspects must be accomplished before they can be transported from the incident. There is also decontamination of tactical gear and weapons as well as law enforcement canines. All of these require special decontamination procedures. Decontamination teams should follow local policies regarding these special decontamination considerations. Hazardous materials responders and local law enforcement should conduct training for these instances.

Remediation of labs

Illicit labs are considered crime scenes and therefore do not fall under the normal hazardous materials scene remediation. Law enforcement ends up being the responsible party to ensure the scene is returned to pre-incident conditions. Many states have programs to assist local law enforcement with lab clean up however proper procedures must be followed to qualify.

Usually law enforcement agencies have private clean up contractors that handle lab clean up. There is also a need for joint training and operations at labs. Joint exercises should be conducted between local fire departments, EMS, local law enforcement and state law enforcement on the hazards present, the operation and remediation of illicit labs. All policies and SOGs should be followed concerning illicit labs. Periodic review of these policies is essential to staying up to date with current methods, hazards and laws.

QUESTIONS

1. List three types of fixed facilities that may house hazardous materials.

2. Where are NFPA 704 placards usually found?

3. The Hazardous Materials Information System (HMIS) marking used by industry is similar to NFPA 704 except for an additional section that was added. What does that section identify?

4. What is the four-digit number placed inside of a DOT placard known as?

5. How many hazard classes are there in the DOT placarding system used for tank trucks, rail cars, and other vehicles transporting hazardous materials?

6. What are the main purposes or goals of the Global Harmonization System (GHS)?

7. What do the green section/pages of the DOT Emergency Response Guide tell first responders?

8. Shipping papers used for rail operations are known as _____ and are kept in either the _____ or the _____.

9. What are the best and most immediate two actions for an awareness-level responder to take upon arrival at a scene?

10. What do the blue section/pages of the DOT Emergency Response Guide tell first responders?

Incidents Involving Terrorism and Weapons of Mass Destruction

This chapter provides knowledge and information related to the following NFPA Standard 1072 JPRs.

5.2.1

6.9.1

OBJECTIVES

Upon completion of this section, you should be able to do the following:

- Identify types of terrorist incidents.
- Identify common terrorist targets.
- Identify indicators of criminal or terrorist activity at biological, radiological and chemical incidents.
- Identify indicators of criminal or terrorist activity at illicit labs.
- Describe signs of exposures to terrorist agents.

INTRODUCTION

The best way to look at this type of incident is that it is the *intentional* release of a hazardous material with intent *to do harm*. Unlike an accidental release, these releases are specifically designed to result in a maximum amount of damage from an explosive device (fig. 5–1) or through the release of a chemical or biological agent. As such, the potential risks involved in responding to these types of incidents increases. The basic step-by-step methodology, however, of analyzing the incident, planning the response, and implementation remains largely the same. Incidents of this nature will likely place a much greater strain on available resources because of the potential for much greater damage and population exposure.

Fig. 5–1. The 1995 bombing of the Alfred Murrah Federal Building in Oklahoma City was designed to create maximum damage to the building. It also took 168 lives.

ACTIONS TO BE TAKEN AT A TERRORIST INCIDENT

In addition to standard precautionary steps taken by first responders at a hazmat incident, a response to a terrorist incident should take the following into consideration:

- Protect yourself and other responders.

- Relay any suspicious or criminal activity to dispatch upon arrival on scene.

- Isolate potentially exposed people or animals.

- Document initial on-scene observations.

- Establish control zones and control access to the scene.

- Take actions to preserve possible evidence.

- Watch for booby traps and/or explosive devices (fig. 5–2).

- Prevent secondary contamination, including from handling patients.

Additionally, because there is a law enforcement component involved with these responses, evidence preservation and documentation become key factors.

Fig. 5–2. Watch for explosive devices at the scene of a terrorist incident. (Courtesy of TEEX)

Potential criminal and terrorist targets

A terrorist/WMD event can occur anywhere. Even so, it is important for the fire department to consider what locations within their communities are most likely to be potential targets. The potential locations can be broken down into several general categories, including occupancy or location, types of events, and significant timing of an event.

The **OTTO** acronym developed by the U.S. Department of Justice can be used to show potential terrorist indicators. The acronym represents the following:

Occupancy and location

Type of event

Timing of the event

On-scene warning signs

Occupancy or location. Certain types of occupancies can have a significant reason for being a target. It may be because of their symbolic or historical value or the purpose for which the building or location is being used. These can be further broken down into the following categories: symbolic and historical, public buildings or assembly areas, controversial businesses, and infrastructure systems.

Symbolic and historical targets represent an organization, idea, or event that is offensive in the mind of the terrorists. These can include structures of historical significance, such as the Washington Monument, Mount Rushmore, the Statue of Liberty, Yankee Stadium, Independence Hall in Philadelphia, or the U.S. Capitol Building (fig. 5–3). The structures can be symbolic, such as the World Trade Center; a building occupied by the Bureau of Alcohol, Tobacco, and Firearms; or the Internal Revenue Service.

Fig. 5–3. Historical buildings or locations can be terrorist targets.

Some terrorists who want to create property damage but not loss of life may strike government buildings such as ROTC offices or military recruiting offices on a weekend when they are typically closed.

Public buildings or assembly areas provide a means of letting the terrorists get attention through the possibility of causing mass casualties. These can include shopping malls, convention centers, entertainment venues, tourist destinations, and places of worship.

Controversial businesses attract the anger of extremist organizations. These can include abortion clinics, animal testing facilities, pharmaceutical companies, nuclear facilities, and furriers. Certain eco-terrorists also target controversial housing developments.

Infrastructure systems are any facilities that are necessary for the continuous operation of society. These include power plants, phone companies, water treatment plants, mass transit, hospitals, data storage facilities, financial institutions, and mass media communication facilities.

Types of events. A terrorist attack occurring at a certain type of event is a possibility. Some common events of concern are large sporting events such as the Super Bowl, the World Series, college and professional basketball playoffs, and racing events that include auto, horse, and marathon races (fig. 5–4). Political events to consider include Republican and Democratic national conventions and rallies for certain political causes. Visiting dignitaries can be another target. When members of the government, such as the president or vice president, attend events in different cities, or when the Pope visits the United States, they may become a target for terrorists.

Fig. 5–4. Popular sporting events can become a terrorist target.

Significant timing of events. Anniversaries of significant events can be triggers for terrorists. April 19 is a trigger for many antigovernment terrorist groups (both the fire at the Branch Davidian Compound in Waco, Texas, and the Oklahoma City bombing occurred on this day). Other event anniversaries can also be triggers, such as the Roe v. Wade anniversary on January 22 or the anniversary of 9/11.

On-scene warning signs. There are numerous signs that would indicate a potential terrorism event. Indicators include explosive devices, chemical labs, indication of biological agents, the release of chemicals into the environment or inside structures, and unusual odors or vapor clouds. Additional signs are addressed throughout this section.

Indicators of a terrorism event

Operations-level responders need to be aware of the signs of different types of terrorist or criminal activities to properly assess the dangers that are present. For example, the ability to differentiate between a chemical and biological release will allow the responder to make better decisions about how to initially handle the situation.

Chemical incidents. **Chemical incidents** are characterized by a rapid onset of medical symptoms, which may take minutes to hours. Indicators may include colored residue, pungent odors, and dead animal and plant life. The following may be observed at a chemical incident scene:

- Unexplained vapor clouds, mists, or plumes

- Abnormal number of sick or dead birds, animals, and/or fish

- Suspicious hazardous materials or lab equipment in the area of contamination

- Intentional release of a hazardous material

- Multiple individuals experiencing unexplained skin, eye, or airway irritation

- Unexplained odors/tastes not normal to the area

- Surfaces with unexplained oily droplets/film

Chemical agents may be present when multiple individuals have the following unexplained health problems:

- Nausea and vomiting

- Twitching, convulsions, or disorientation

- Sweating

- Pinpoint pupils or runny nose

- Tightness in chest, difficulty breathing

- Death

Biological agents. In cases of **biological incidents**, there are typically no indicators found because biological agents are generally colorless and odorless. In such incidents, the onset of symptoms may take days to weeks, depending on the agent involved. As a result of the delayed onset of symptoms, an affected person has a greater chance of transmitting the disease to others. The migration of these infected individuals may increase the area of the incident dramatically. Possible indicators of a biological release include:

- Multiple casualties with similar signs and symptoms reported by health care agencies

- Unscheduled or unusual spraying, especially at night

- Abandoned spray devices

- Unusual increase in the number of sick or dying people or animals

Radiological. The detonation of a large nuclear device or weapon would be readily apparent. Smaller or improvised devices may initially appear to be a very large conventional explosive device. Responders will not know if the device was radiological in nature until monitoring is completed. Some questions to consider when making the determination are:

- Was there a threat given?

- Is the area a potential target?

- Are there any suspicious packages, vehicles, or activities?

- Was there an explosion?

Illicit laboratories. Illicit labs may be used for the development of illegal drugs, weapons production, or WMD (fig. 5–5). When trying to identify an illicit lab, always look for indicators that do not fit the surroundings. Illicit labs can be found in the trunks of cars, motel rooms, trailer parks, hotel rooms, and self-storage buildings. Possible indicators include:

- Suspicious odors (pungent, sweet, bitter)

- Explosions/fires

- Blacked-out windows

- Unusual ventilation and drainage systems

- Stained soil

- Dead trees/vegetation

- Burn pits

- Propane tanks with corroded fittings

- Solvents (hexane, acetone, etc.)

- Multiple containers of over-the-counter medications

- Large quantities of hazardous materials not typical for the area

- Laboratory equipment
  - Beakers, tubing, and burners
  - Two-liter bottles, mason jars with tubing attached
  - Hot plates, electric skillets, slow cookers
  - Hand scales

- Bomb making equipment

Fig. 5—5. Illicit laboratories can be used to make illegal drugs or terrorist weapons.

Explosives. The signs of an explosion are obvious to response personnel. The determination that must be made is whether it was done intentionally. The following are some questions to consider:

- Was it a potential or logical target?

- Was there a threat given?

- Were there any secondary explosions?

- Were any pre-detonation indicators observed by witnesses?

In the event that personnel are on scene prior to the detonation, they should be on the lookout for suspicious packages that are left unattended, shipped or mailed with no return address, are misshapen or stained with an oily residue, or have protruding wires. They should also be aware of any suspicious activities such as people who are out of place, appear to be aggressive, or are taking photos or video of a potential target. Additional clues are suspicious vehicles parked in front of buildings, at hazmat storage areas, or in restricted areas. Unfamiliar vehicles loitering for long periods of time is also considered suspicious activity. The first responder should also note any type of device attached to containers such as storage tanks, pipelines, flammable liquid or chemical containers, and compressed gas cylinders.

DOT classification of WMD agents

Terrorists use many different weapons and means to achieve their goals. Some of them are very subtle; others meet their goal with extreme violence. Although WMD agents are often thought of as having the specific purpose of injuring or killing, therefore being outside of the standard DOT hazmat classification system, this is not the case. Every type of device or substance used as a warfare agent has a place in the DOT hazard classification system, allowing first responders to use many of the same informational resources to determine how to handle the response. However, after it is determined that a warfare agent is involved, it is always a good idea to seek out resources with specific information on that substance. Warfare agents include the following:

- Chemical agents
  - Nerve agents
  - Vesicants (blister agents)
  - Blood agents
  - Choking agents
  - Irritants (riot control agents)
- Biological agents and toxins
- Radiological materials

These weapons have been grouped into categories with similar properties. It is important to note that some weapons or agents do not easily fall into a single category. There are two common acronyms used to explain weapons of mass destruction. The first is CBRNE (chemical, biological, radiological, nuclear, and explosive); the other is B-NICE (biological, nuclear, incendiary, chemical, explosive). The major difference between the two is that B-NICE separates incendiary and explosives, whereas CBRNE does not. This section discusses the acronym B-NICE.

Biological agents and toxins. Biological agents and toxins are Hazard Class 6 toxic and infectious substances. They are most likely to be used in a terrorist incident because they are the easiest for a terrorist to produce.

Examples of biological agents include the following:

- Anthrax (DOT 6.2)
- Mycotoxins (DOT 6.1 or 6.2)
- Plague (DOT 6.2)
- Tularemia (DOT 6.2)
- Ricin (DOT 6.1)

Routes of entry depend on the compound and include skin contact, inhalation, or injection.

Anthrax is a naturally occurring bacteria found in dead sheep (fig. 5–6). Exposure occurs through skin contact or inhalation of the anthrax spores. Anthrax is relatively easy to obtain; however, the military-grade fatal type of anthrax is difficult to culture and produce. It must be distributed under certain conditions to be effective.

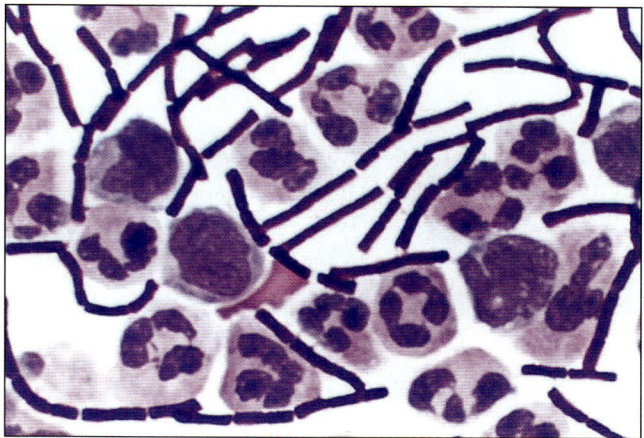

Fig. 5–6. Anthrax falls under Hazard Class 6.

Ricin is a widely available toxin that is easily produced. It is 10,000 times more toxic than sarin, a chemical weapon. However, unlike sarin, ricin is difficult to distribute and must be injected to be effective. People exposed to ricin via inhalation or ingestion will most likely become sick and recover. It is popular among domestic terrorists, second only to explosives.

Exposure and its effects. Effects of these agents on the body vary widely depending on the agent involved. In many cases individuals present with flu-like symptoms that get progressively worse. Hemorrhaging and the shutdown of various body systems can occur. In some instances, rashes on the skin are evident.

Detection. On-site detection methods at this time are very limited and in many cases not available at all.

Nuclear (radiological) agents. Radioactive agents fall under Hazard Class 7. There are three primary methods a terrorist group could use to distribute nuclear material in an attack. The first is to detonate a nuclear fission device (weapons grade). This method causes the most damage but is least likely to occur due to the difficulty in obtaining the device.

The second method would be to create a **dirty bomb** by using an explosive to scatter radioactive material. This would result in damage from the initial explosion as well as from the short- and long-term effects from the radioactive material.

The last method would be to target any type of facility that uses or stores radioactive material, such as a weapons facility, nuclear power plants, and so on (fig. 5–7).

Fig. 5–7. Nuclear power plants can be targeted by terrorists to create a radioactive release.

Exposure and its effects. These effects would be the same as discussed in chapter 3 regarding exposure to radiation.

Detection. Radiological material can only be detected by equipment such as a Geiger counter.

Incendiary devices.

Incendiary devices fall under Hazard Class 3 flammable liquids and have been used by terrorist organizations for centuries. In fact, they are probably the easiest to obtain materials for and to build. These devices have been very popular in Europe for decades and have been on the rise in the United States since the mid-1980s. The Federal Bureau of Investigation (FBI) Bomb Data Center says that incendiary devices are used in 20% to 25% of all bombing incidents in the United States and are successfully ignited about 75% of the time. One of the major problems for investigators is determining whether an incendiary device was used for the purposes of terrorism or arson.

Signs indicating the use of incendiary devices for a terrorist event are similar to those found at a suspicious or arson fire. It is important to note that these following signs are not a guarantee that an incendiary terrorist event has taken place, but they are signals for the firefighter to be diligent and take precautions while operating at the incident. These include multiple points of fire origin, signs of accelerant use, containers of flammable liquids, splatter patterns indicating a thrown device, residue from the fusing material, signs of forced entry to the structure, common appliances out of place for the location, and so on (fig. 5–8).

Fig. 5–8. Containers of flammable liquids can be a sign of accelerant use.

Methods of detection. The use of incendiary materials can be detected using several different metering technologies, such as colorimetric tubes or combustible gas meters.

Means of protection. When dealing with incidents involving incendiary devices, firefighters should wear full PPE including SCBA. Avoid handling the device, and keep clear of any vapors, mists, or liquids until it is determined what they are. If unsure of the nature of the device, technical assistance should be called. To protect responders and civilians involved in the incident, proper decontamination procedures should be implemented prior to transporting to a hospital.

Chemical agents.

Chemical agents fall under a broad category of materials that can be broken down into several types based on their composition and their effects on the body. These substances range from materials specifically designed to be used in warfare to those used in industrial applications.

Nerve agents.

The following nerve agents are related to organophosphorus pesticides and fall under DOT Hazard Class 6.1. The two-letter abbreviations following each compound represents its military designation.

- Tabun (GA)
- Sarin (GB)
- Soman (GD)
- V agent (VX)

These nerve agents were designed for the sole purpose of killing people. The ability to kill large numbers of people is dependent on the dispersion device. To produce the desired effect, the nerve agent must be either touched in its liquid form or inhaled in its aerosol form. These materials do not remain in vapor form very long. Military warfare agents have vapor pressures lower than water. Vapors are not created from a pool of liquid, and it is generally not a major hazard unless the liquid form is touched or placed on the skin.

Exposure and its effects. Exposure to these agents can be via inhalation or skin contact with a liquid. The inhalation of a small amount can cause pinpoint pupils, runny nose, and mild difficulty breathing. If a larger amount is inhaled, it can cause sudden loss of consciousness, convulsions, temporary breathing stoppage, flaccid paralysis, copious secretions, and death.

Symptoms from exposure to a liquid type can range from localized sweating, nausea, vomiting, and a feeling of weakness, to a sudden loss of consciousness, seizures, breathing stoppage, copious secretions, paralysis, and death (table 5–1). Some of the warning signs that exposure has occurred are runny nose, difficulty breathing, and uncontrolled muscles and bodily functions. A fruity odor may also be reported. An easy way to remember the signs and symptoms of an exposure to nerve agents is to use the acronym SLUDGEM:

- **S**alivation
- **L**acrimation (watering of the eyes)
- **U**rination
- **D**efecation
- **G**astrointestinal upset
- **E**mesis (vomiting)
- **M**iosis (constricted pupils)

Table 5–1. Nerve agents—symptoms of exposure

Central Nervous System	Autonomic Nervous System	Neuromuscular
Respiratory arrest	Sweating	Weakness
Disorientation	Diarrhea	Trembling
Slurred speech	Nausea	Paralysis
Depression	Abdominal pain	Respiratory failure
Respiratory depression	Vomiting	
Headache	Reduced vision	
Convulsions	Pinpointed pupils	
Coma	Drooling	

Methods of detection. Some common methods of detection are detection papers such as M8 or M9, colorimetric tubes, military detection kits, pesticide tickets, and electronic meters.

Means of protection. If during a response to an incident it is suspected that a nerve agent has been used, it is best to follow the procedures found in the **North American Emergency Response Guide (NAERG)**, Guide 153. Common antidotes are atropine and 2-PAM chloride.

Blister Agents (vesicants).
Blister agents are designed to incapacitate the enemy. These compounds fall under DOT Hazard Class 6.1. Although these agents are toxic at high concentrations, the primary hazard is skin contact resulting in blistering and severe irritation. Blister agents pose less of a hazard than nerve agents.

Examples of blister agents include:

- Mustard agent (H)
- Distilled mustard (HD)
- Nitrogen mustard (HN)
- Lewsite (L)

A military weapon, blister agents are designed to incapacitate the enemy in two ways:

- The exposed person would be overcome with severe irritation and blisters of the skin (fig. 5–9).
- Several other people would be required to help treat the victim, tying them up and preventing them from fighting.

Exposure effects can be delayed from 15 minutes to several hours; quick identification is the key to patient treatment.

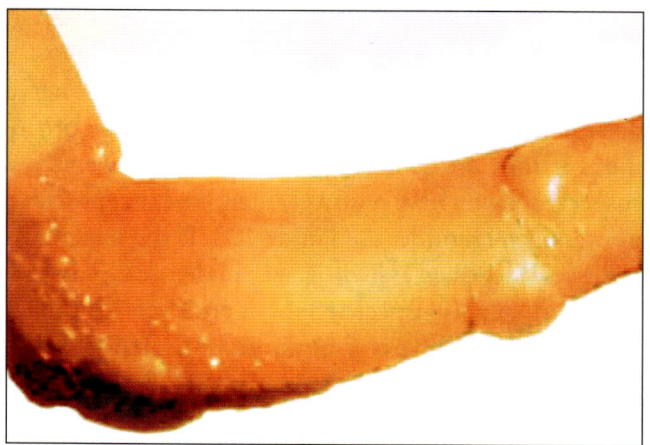

Fig. 5–9. Blister agents cause severe irritation and blistering of the skin. (Courtesy of the CDC)

Incidents Involving Terrorism and Weapons of Mass Destruction

HAZMAT

Chapter 5

Exposure and its effects. Vesicants can contact a person either as a liquid or a vapor. Vapors are more easily developed in warmer climates. Their primary effect occurs on the skin, the respiratory tract, the eyes, and the gastrointestinal tract. The mildest form is a reddening of the skin, similar to a sunburn with an itching or burning sensation. Blistering will occur later, and it can act like an acid on unprotected skin. These agents can cause respiratory failure, which can lead to death. Eyes are the most sensitive to vesicants. It takes less time from exposure to injury with the eyes than with the skin. The least affected is the gastrointestinal tract.

Its impact on the central nervous system seems to be minimal when persons are exposed to small amounts. Studies have shown that exposed persons appear to be sluggish, apathetic, and lethargic.

Detection methods. Similar to nerve agents; self-protection procedures can be found in the NAERG, Guide 153.

Blood agents. Blood agents, otherwise known as chemical asphyxiants, disrupt the blood's ability to transport oxygen. Examples of blood agents include:

- Hydrogen cyanide (AC) DOT 6.1

- Cyanogen chloride (CK) DOT 2.3

Exposure and its effects. Contact with agents can be either from a vapor or a liquid. Liquids will vaporize quickly due to the chemical's volatility. These cyanides react with the hemoglobin in the blood and prevent it from carrying oxygen throughout the body. Its effect on the body is the same as asphyxiation, but more rapid. High concentration can lead to respiratory and cardiac arrest. Common signs that exposure has occurred are great difficulty in breathing and onset of cardiac symptoms. Some victims may report an odor of bitter or burnt almonds.

Detection methods. Similar to those used for nerve agents; firefighters should use the following procedures if they are at an incident and blood agents are suspected. If you have positively identified the substance as cyanogen chloride, use Guide 125. If the material is positively identified as hydrogen cyanide, use Guide 117. If a blood agent is suspected but not positively identified, use Guide 123. There is an antidote kit for blood agents called the Pasadena Cyanide Antidote.

Choking agents. Choking agents are common in industrial applications. They fall under DOT hazard class 2.3. One example is chlorine, which is used in water

treatment facilities and swimming pools, thus making it present in almost any community.

Examples of choking agents include:

- Chlorine (CL) DOT 2.3

- Phosgene (CG) DOT 2.3

Exposure and its effects. Exposure to these chemicals is usually through inhaled vapors. The primary effect is pulmonary edema, or when the lungs fill with fluid and severe pneumonia sets in. Symptoms of this are eye and airway irritation, dyspnea, chest tightness, and delayed pulmonary edema. Pulmonary distress among victims is an outward warning sign. Other signs are odors such as chlorine bleach or swimming pool odors and the smell of newly mown hay or grass, which is an indicator of phosgene.

Detection methods. Similar to other types of chemical agents; operating procedures when these chemicals are discovered can be found in NAERG Guide 124 for chlorine and Guide 125 for phosgene. If the material has not been determined then the procedures found in Guide 123 should be followed.

Irritants. **Irritants** are the most commonly used of all the potential terrorism agents. They are not extremely toxic. Symptoms generally dissipate within 15–20 minutes after the victim is exposed to fresh air.

Examples of irritants and their respective DOT class are as follows:

- Mace (CN) DOT 6.1

- Tear gas (CS) DOT 6.1

- Pepper spray (OC) DOT 2.2

As a point of consideration, pepper spray falls under DOT hazard classification of 2.2, but its subsequent hazard risk is 6.1.

Exposure and its effects. These agents cause discomfort and eye closure (fig. 5–10). They tend to render the recipient temporarily incapacitated. The major reason for their use is to cause pain, burning, or discomfort on exposed mucous membranes and skin. This occurs within minutes of being exposed. Initial signs of exposure are the smell of hair spray and pepper.

Detection methods. Detection of these sprays can be made through the use of laboratory testing using gas chromatography/mass spectrometry. Self-protection is accomplished by following the procedures in the NAERG Guide 159 if the material has been identified; if not, use

Guide 153. For any victims affected by these agents, have them decontaminated and transported to a hospital.

Fig. 5–10. Pepper spray causes discomfort and eye closure. (Courtesy of the DOD)

Figure 5–11 shows the relative lethality of selected agents.

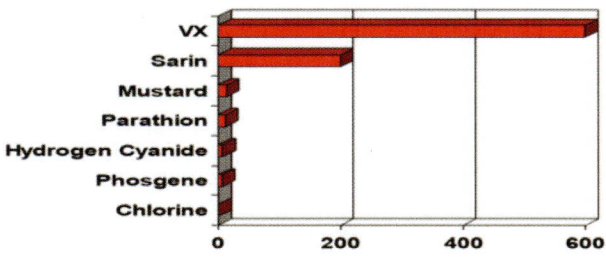

Fig. 5–11. Relative lethality of agents

Explosives. Bombs appear to be the weapon of choice for terrorists. Approximately 70% of all terrorist incidents involve the use of explosives. FBI bomb data reports that, on average, there are 3,000 bombings per year in the United States. The majority of these incidents are small bombings; however, the data shows that these explosions kill an average of 32 people per year and injure 277.

Improvised explosive devices (IED) can be designed to deliver an assortment of harm and destruction, and can also be used as a means of dispersing chemical, biological, incendiary, and nuclear agents (fig. 5–12). Explosives are defined as materials capable of violent decomposition. This decomposition often takes the form of extremely rapid oxidation. Explosions are the result of a sudden and violent release of gas during the decomposition of explosive substances. This release is followed by high temperature, strong shock, and loud noise.

Fig. 5–12. IEDs can be designed to deliver an assortment of harm and destruction.

Scene safety

The first responder should consider the following safety issues when making an entry:

- There may be secondary events designed to incapacitate emergency responders.

- There is the potential for armed resistance and use of weapons.

- There could possibly be booby traps.

- There may be secondary contamination from handling patients.

It must be considered a possibility that the individual(s) who released the material or detonated the device will remain in the area to watch or plan to hinder emergency operations. Law enforcement may need to clear the area of any potential threats before any other operations can begin. There is also the potential for secondary explosive devices to be set in place with the intent to injure onlookers or emergency responders and further increase the chaos level. These devices can be triggered either by a remote device or by some type of a trip switch. Bomb squad personnel should be called in to ensure that there are no additional active devices before operations proceed.

Evidence preservation

It is vital that emergency responders consider the preservation of evidence. To properly preserve and collect evidence, the following actions should be taken:

- Do not move or touch anything that appears to be evidence unless absolutely necessary.

- Gently cover evidence that may be damaged by the weather or other external sources.

- Document all actions (fig. 5–13).

Fig. 5–13. Personnel operating at the scene of a terrorist incident should document all actions taken.

The collection of evidence should be conducted by law enforcement personnel only. Hazmat response personnel should collect evidence only if they have received proper training or if required to do so by law-enforcement personnel. It is vital to maintain a chain-of-custody record whenever it becomes necessary to collect or transport possible criminal evidence.

Additional assistance resources

The first responder should review the procedures outlined in SOPs/SOGs and LERP for requesting state, federal, and military assistance at a terrorist incident. These incidents require a strong law enforcement component for crime scene preservation and security as well as evidence preservation.

At the federal/military level, the following resources are available for additional assistance.

FBI Hazardous Materials Response Unit (HMRU).
The HMRU can provide assistance with the following:

- Identification

- Mitigation

- Evidence collection

Army's Technical Escort Unit (TEU).
The TEU can provide assistance with the following:

- Identification

- Mitigation

- Assistance with chemical, biological, and explosive materials

Marine Corps Chemical Biological Incident Response Force (CBIRF).
The CBIRF responds across the country and around the world to terrorist acts. It consists of three main sections:

- Decontamination

- Security

- Medical

This response force can also provide detection and mitigation assistance.

CONCLUSION

Upon arrival at a hazmat incident, controlling the scene to prevent additional exposures is a priority task. As this is being done, the incident command system needs to be established to better organize and implement the response. After sufficient information has been collected regarding the product release, the incident commander can set the response objectives for the operation. These objectives must be realistic in nature based on the available resources, and must be specific and measurable. After the product hazards have been identified and the response objectives defined, operational personnel can determine the appropriate PPE needed to meet those objectives. Lastly, although in many ways it is handled in the same basic manner as an accidental release, incidents involving criminal or terrorist activities require addressing additional concerns in the planning phase to ensure the safety of responders and civilians alike.

QUESTIONS

1. Name three reasons that a hazmat responder would choose to protect-in-place rather than evacuate.

2. In which control zone is the incident command center located?

3. Name the positions and groups or branches that are special to a hazardous response Incident Command System (ICS):

4. What are the three hazardous materials incident levels?

5. List the three different response actions.

6. List the six strategic priorities at a hazardous materials/WMD incident.

7. Which type of SCBA provides the highest level of respiratory protection in a hazardous materials incident?

8. Describe the four levels established by the Environmental Protection Agency for chemical protective equipment.

9. The process by which a hazardous material moves through PPE material at the molecular level is called:

10. Responders might choose structural firefighting clothing for what type of hazard?

11. Name five additional response actions that should be taken when a hazardous materials incident is the result of a terrorist act.

12. The U.S. Department of Justice acronym OTTO stands for what potential terrorism indicators?

13. What type of terrorist incidents are characterized by a rapid onset of medical symptoms?

14. Name five possible signs that an incendiary device has been used to start a suspicious fire.

15. The acronym SLUDGEM is an easy what to remember the symptoms of exposure for what type of mass destruction agent? What does the acronym stand for?

Chemicals and Their Properties

This chapter does not directly correlate to the NFPA 1072 standard. However, the information contained herein is valuable to the first responder and should be studied in order to gain a better understanding of hazardous materials/WMD response.

OBJECTIVES

Upon completion of the chapter, you should be able to do the following:

- Describe the different classes of hazardous materials as defined by the Department of Transportation (DOT)
- Understand the basic chemical properties of the different hazmat classes
- Have a basic understanding of the periodic table of the elements
- Have a basic understanding of the elements of a hazmat/chemical reaction
- Have an understanding of basic chemistry

INTRODUCTION

There is no shortcut to learning chemistry, even the abbreviated form which follows. To learn to read chemical symbols and molecular formulas, you should memorize the chemical symbols of 50 or more elements. This can be confusing, but it is certainly not difficult. All of the elements' symbols are related to their names. The confusing part is that the symbols of some elements come from their Latin, Greek, or other foreign language name, such as Na for sodium (Latin: natrium) or K for potassium (Latin: kalium).

In all cases, the chemical symbol for the element is one or two letters, with the single letter or the first of two letters used always capitalized, and the second letter always lowercase. The first letter is always the first letter of the name (remember that it may not be the English name), and the second letter of the symbol may be the second or third (or more) letter of the name.

Note that the shorthand notation for an element is called the chemical symbol for the element, but when a compound is being discussed, its shorthand notation is called a chemical formula. The chemical symbol stands for the name of the element, and for one atom of the element, while the chemical formula shows the elements and how many atoms of each element are in a particular compound. Thus, the symbol for elemental oxygen is O, while the formula for molecular oxygen (the form in which oxygen exists in the atmosphere) is O_2. The subscript 2 indicates there are two atoms of oxygen present in molecular oxygen, just as the chemical formula O_3 stands for ozone, the 3 indicating that there are three atoms of oxygen present in the compound ozone. The formula H_2SO_4 states that there are two atoms of hydrogen, one atom of sulfur, and four atoms of oxygen in the chemical compound sulfuric acid.

BASIC CHEMISTRY CONCEPTS

Chemistry is universally defined as the science of matter, energy, and reactions. A working fire is a perfect example of chemistry at work. Matter is defined as anything that occupies space and has mass. Matter is made up of atoms that combine in various ways to form compounds. The material that is burning is matter, whether it is wood, paper, gasoline, coal, natural gas, or anything else that burns. Matter may be solid, liquid, or gas. Matter comes in nonflammable and noncombustible forms, also, so the matter that burns is a fuel. Indeed, the definition of a fuel is anything that will burn.

The matter that is supporting the combustion of the fuel is an oxidizer. An oxidizer is a substance that either is oxygen, a substance that contains oxygen and gives it up readily, or another substance that will support combustion. The products of combustion (in the case of hydrocarbons) carbon, carbon monoxide, carbon dioxide, water, and several intermediate products are all matter. Therefore, matter is always involved in a chemical reaction, either as the original material(s), or as the product(s) of the reaction. A chemical reaction is a chemical change that occurs when two or more substances are brought together and energy is either absorbed or liberated. Chemical reactions include oxidation (combustion is a form of an oxidation reaction), condensation, hydrolysis, ionization, photosynthesis, polymerization, pyrolysis, and reduction.

Energy is defined the capacity for doing work. Heat and light are forms of energy. Energy may be generated by electrical, mechanical, or chemical means.

Elements are the basic building blocks of the universe. They are pure substances that cannot be broken down by chemical means. There are 112 elements known and they are listed on the periodic table of the elements in a particular manner. The first 92 elements are naturally occurring, and elements with atomic numbers of 93 and higher are synthetic. A few elements exist freely in nature, but most are bound chemically in the chemical compounds that make up the composition of the plant and animal life and minerals that make up the Earth, the water that covers three-fourths of it, and the atmosphere that surrounds it. They also make up all the stars and other planets in the universe, from which the naturally occurring elements probably originated.

Atoms are the smallest particles of an element that can still be identified as that element, and they are electrically neutral (i.e., there is no electrical charge). No matter what is done to atoms chemically, they cannot be broken down into simpler particles and still be identified as that original atom. Atoms, however, can be broken up by physical means. The nucleus of certain atoms can be broken up by the bombardment of nuclear particles from other atoms. Atoms are made up of two different subatomic particles, the nucleus and the electrons. The nucleus is the subatomic particle at the center of the atom. It has all the positive charge of the atom, and essentially all the weight of the atom. The nucleus contains two nuclear particles, the proton and the neutron. The proton is a nuclear particle that has an atomic weight of 1 atomic mass unit (AMU) and an electrical charge of positive 1 (+1). The other particle inside the nucleus is the neutron, a nuclear particle that weighs 1 AMU and has no electrical charge. Thus the nucleus contains all the positive charge of the atom and essentially all its weight. Spinning around the nucleus (at the speed of light) are the electrons, which are subatomic particles that have an electrical charge of negative 1 (−1), and essentially no weight. As stated above, the atom has been found to be electrically neutral, which means that there must be the same number of electrons in orbit around the nucleus as there are protons in the nucleus.

The elements are arranged on the periodic table according to their atomic number, which is the whole number found above the symbol for the element in its box on the periodic table. Sometimes it is found to the left above the symbol, sometimes to the right above the symbol, and on other tables, it may be directly above the symbol. Even if it appears below the symbol (rarely), it will always be a whole number, never in parentheses. The decimal number that appears below the symbol (sometimes it appears above) is the atomic weight or atomic mass of the element. The atomic number of an element is equal to the number of protons in each atom of the element. Therefore, oxygen, atomic number 8 on the periodic table, has eight protons in the nucleus of every atom of oxygen in the universe. The number of protons in the nucleus (the atomic number) defines the element. That is, wherever an atom is found to have eight protons in its nucleus, by definition it is an atom of oxygen.

The **atomic weight** (or alternatively, **atomic mass**) of an element is the combined weight of the total number of protons and neutrons in the nucleus of that element (remember that the electrons in orbit are essentially weightless). It is given as a decimal number because, in reality, the atomic weight is a weighted average of all the isotopes of the element. An isotope is a form of the same element, having identical chemical properties (because of the same number of protons), but a different number of neutrons in the nucleus of its atoms.

THE PERIODIC TABLE OF THE ELEMENTS

There were many attempts by early chemists to try to organize the known elements into some system where they could be grouped together by chemical properties. It was known that several elements had similar chemical properties, but that they had differing atomic weights. By the middle of the 19th century, it was believed that if the elements were arranged in order of their increasing atomic weights. In 1869, Dmitri Mendeleev discovered that by arranging the elements in order of increasing atomic weight, he could arrange the elements together in groups (families) and periods. That is, every so often (periodically) an element would repeat the chemical properties of a lighter element. He constructed a periodic table of the elements in which he left spaces for elements he inferred existed but had not yet been discovered. As proof that he was right, there were, indeed, elements discovered that fit in the spaces Mendeleev left on his table, and they had the chemical properties he predicted they would have.

The proper arrangement for elements on the periodic table is by atomic number, not atomic weight. Nonetheless, this correction produced only very small changes in the table that we know today (fig. 6–1).

The periodic table of the elements today comes in many forms, all very similar, with the main differences being the headings at the top of each group and the placement of the numbers representing the atomic numbers and atomic weights. Some tables include the name of the element in addition to the symbol, and some tables include the number of isotopes and the electronic configuration. The table used in this chapter is very simple, containing only the symbol (which stands for the element and also represents an atom of that element), the atomic number (to the top left of the symbol), and its atomic weight (below the symbol). Where the atomic weight is shown in parentheses, it represents the atomic weight of the most stable isotope of the element.

Dividing the table of the elements

The periodic table can be divided into two types of elements: metals and nonmetals. A line can be drawn below hydrogen in Group IA, then stepwise below boron in Group IIIA, between silicon and aluminum, under silicon, between arsenic and germanium, under arsenic, between tellurium and antimony, under tellurium, and between astatine and polonium. All elements to the left and below the line are metals, while all elements above and to the right of the line are nonmetals. These two types of elements react differently and form different types of compounds, which will be discussed later. For now, it is very important to know the different kinds of elements and where they are on the periodic table. Not all tables have this line. There are other divisions within the periodic table. The tall columns marked IA through VIIIA are called the representative or main group elements, made up of both metals and nonmetals. The short columns marked IB through VIIB and VIII are called the transition elements and are all metals. At the bottom of the table are two series of metals, called the lanthanide series and the actinide series, collectively called the rare earth metals. It is obvious that the vast majority of the elements are metals.

There are four out of eight groups of elements in the vertical columns (the representative elements) that demonstrate strong similarities in their chemistry. They are the Group IA elements, called alkali metals, headed by lithium (even though hydrogen sits above lithium); Group IIA, the alkaline earth metals, headed by beryllium; Group VIIA, the halogens, headed by fluorine; and Group VIIIA, the noble or inert gases, headed by helium. These groups all exhibit the family effect, the phenomenon that all the elements in the group or family have similar chemical properties. It can be seen that the family effect is strongest on the edges of the periodic table, and it weakens toward the middle of the table. The family effect is caused by similarities in the electronic configuration of the atoms of the elements. The chemistry of every atom is due to the configuration (or positions) of the electrons in different orbits around the nucleus of each atom. All the atoms of an element have the same number of electrons, and they are all arranged in the same manner. Remember that since all atoms are electrically neutral, there must be the same number of electrons in orbit around the nucleus as there are protons in the nucleus. We also know how many protons are in the nucleus of each atom of a particular element by looking at the periodic table and noting the atomic number of the element.

The family effect

All the elements within the main groups on the periodic table of the elements have the same chemical characteristics as other elements in the same group. This is known as the family effect, and it is strongest at the edges of the table and weakest toward the center of the table. This is because the closer an element is to stability, the stronger its propensity for giving up, accepting, or sharing electrons. An important tip to remember is that all of the representative elements have the same number of electrons in their outer ring (also called shell, orbital, or valence ring) as the number of the group they are in. The family effect of certain main groups is the proof of the statement that the chemistry of an element is a consequence of its electronic configuration, especially the electrons in the outer or valence ring of the representative elements. The family effect is the why of chemical reactions of the elements.

Elements important in hazardous materials

It is very important to be familiar with the periodic table to understand how elements react with one another to form chemical compounds. It is not necessary to memorize the chemical symbols of all the elements, but one must be familiar with those that most often appear in hazardous materials. Many other elements, like gold (Au), silver (Ag), and platinum (Pt) are important in commerce, but are not commonly found in compounds that will be involved in hazardous materials incidents (there are a few hazardous silver compounds).

As stated previously, the metallic elements will always give up electrons to reach stability, and the nonmetallic elements will either accept electrons or share them to complete their outer ring. The giving up, accepting, or sharing of electrons is done through the formation of chemical compounds. A chemical compound is defined as the chemical combination of two or more atoms, either from the same element or from different elements, that is electrically neutral. There are two types of chemical compounds important to the student of hazardous materials. They are ionic compounds and covalent compounds. There are other types of chemical compounds that exist, but they are not as important to this study as ionic and covalent compounds.

HYDROCARBONS

Hydrocarbons are defined as covalent compounds containing only hydrogen and carbon. They are important in the study of hazardous materials because all hydrocarbons will burn. More than 90% of hazardous materials calls around the country involve a hydrocarbon material such as gasoline, kerosene, and fuel oil. Some will burn explosively, and some more slowly, but they will all burn. Carbon's unique structure makes it the basis of organic chemistry, which is the study of compounds that are or were once a part of living things. The way carbon combines with hydrogen and other nonmetals in covalent compounds makes it one of the most important elements for life and for the study of hazardous materials. Carbon not only combines covalently with other nonmetals but also with itself. This fact may not seem remarkable, considering that oxygen reacts with itself to form O_2, hydrogen reacts with itself to form H_2, and nitrogen reacts with itself to form N_2. These molecules, which are made up of two atoms bonded together are therefore diatomic molecules. The ability of the elemental gases to combine with themselves is limited to the formation of these diatomic molecules. Oxygen can actually form a triatomic molecule with itself called ozone, O_3. However, carbon has the ability to combine with itself almost infinitely, but not as a molecule containing only one element, as the gases form.

Cyclical hydrocarbons: the aromatics

The simplest, most common, and most important hydrocarbon in this group is called **benzene**. It is the first of the six-carbon cyclical hydrocarbons called aromatic hydrocarbons. Because of the major differences in the structural formulas, there are some corresponding major differences in properties between and among benzene and the straight-chain hydrocarbons of the same carbon content. This particular hexagonal structure is found throughout nature in many forms, almost always in a more complicated way, usually connected to many other "benzene rings" to form many exotic compounds, most of which are not really understood. We will, however, be concerned with just benzene and a few of its derivatives. Those hydrocarbon compounds include toluene and xylene.

There are other cyclical hydrocarbons, but they do not have the structural formulas of the aromatics, unless they are benzene-based. These cyclical hydrocarbons may have three, four, five, or seven carbons in the cyclical structure, in addition to the six-carbon ring of the aromatics. None of them has the stability or the chemical properties of the aromatics.

Naming hydrocarbons and hydrocarbon derivatives

There exists an agreed-upon system for naming the hydrocarbons and their **derivatives**. All hydrocarbons are relatively stable; that is, they do not enter into controlled chemical reactions readily (fire is usually an uncontrolled chemical reaction). For the hydrocarbon to react, a reactive site on the hydrocarbon chain must be created. This is done by the chemist or chemical engineer removing a hydrogen atom from the chain, thus creating two radicals, the hydrogen atom with one unpaired electron and the resulting hydrocarbon chain with one unpaired electron.

Flammable liquids

Some hydrocarbons, hydrocarbon mixtures, and hydrocarbon derivatives are flammable liquids. The shorter-chain alkanes, methane, ethane, propane, and butane are all gases, but the next five hydrocarbons, pentane, hexane, heptane, octane, and nonane, and their branched isomers, are all flammable liquids. The same is true for the alkenes, with ethene, propene, butene, and their isomers being gases. The next five alkenes, 1-pentene, 1-hexene, 1-heptene, 1-octene, and 1-nonene, and their isomers, are all flammable liquids. The "1-" in front of the name indicates the position of the double bond, being attached to the first carbon in the chain. In the cycloalkanes, cyclopropane and cyclobutane are gases, whereas cyclopentane, cyclohexane, cycloheptane, and cyclooctane are flammable liquids. The simplest aromatic hydrocarbons are flammable liquids, benzene, toluene, and the xylenes. The general rule in an analogous series of compounds is that those with the lightest in molecular weight are gases, the next heaviest are liquids, and the rest are solids. This is called the **weight effect** of molecules. The alkanes, alkenes, and cyclical and aromatic hydrocarbons (and all their derivatives that follow the flammable liquids in the length of the chain) are thus increasing in molecular weight. These are followed by the heaviest compounds in each series, which are all solids.

The most common flammable liquid in commerce today is a mixture of hydrocarbons from pentane to nonane (including isomers), plus additives. It is gasoline, and it is usually formulated to have a flash point of −40°F. There are other mixtures of alkanes and other hydrocarbons, usually the liquids that are left after petroleum "cracking" and distillation, that are flammable liquids, but these cost too much to separate into individual compounds. These mixtures are flammable liquids that find uses as solvents and cleaners, but still present a hazard as flammable liquids. The names include naphtha, Stoddard solvent, lighter fluid, petroleum ether, petroleum spirits, and many others. The 10 carbon alkanes and alkenes (and other hydrocarbon series) begin the combustible liquid regime, up to the solid hydrocarbons, which are all combustible.

Halogenated hydrocarbons include more compounds than just the alkyl halides (halogenated derivatives of the alkanes and alkenes). They include any compound that started as a hydrocarbon and has had one or more halogen atoms substituted for hydrogen atoms. This means that a certain halogenated hydrocarbon may have as its hydrocarbon backbone a radical formed from an alkane, an alkene, a diene (an unsaturated hydrocarbon with two double bonds), a cycloalkane, an aromatic hydrocarbon, or, indeed, any other hydrocarbon imaginable. Halogenated hydrocarbons run the gamut from flammable liquids, to combustible liquids, to liquids that are almost impossible to ignite, to fire-extinguishing agents. An additional hazard of the halogenated hydrocarbons is that their vapors are harmful to breathe, causing problems from irritation of the respiratory system to possible carcinogenicity.

Alcohols are a group of hydrocarbon derivative compounds characterized by the presence of the hydroxyl radical attached to a hydrocarbon backbone, as their general formula, R−OH, shows. The alcohol series has no gases; even the lightest, methyl alcohol, is a liquid. The lightest in the series are all flammable liquids, including methyl alcohol, ethyl alcohol, propyl and isopropyl alcohols, butyl alcohol and its isomers, amyl alcohol and its isomers, denatured alcohol, methallyl alcohol, propargyl alcohol, crotonyl alcohol, and allyl alcohol. The longer-chain liquids are combustible liquids, and the heaviest alcohols are combustible solids. The proper names of the alcohols now end in ol, which means methyl alcohol is now methanol, ethyl alcohol is now ethanol, and so on.

The **ethers** are an especially dangerous series because of their volatility, reactivity, and tendency to form explosive peroxides when exposed to air. The common ethers that are flammable liquids include allyl ether, dibutyl ether, diethyl ether, divinyl ether, ethyl butyl ether, methyl ethyl ether, and many others. Any ether should be considered a flammable liquid until proven otherwise (it will still be combustible). Care must be taken to prevent air from reaching the ether molecules.

The **ketones** are a class of hydrocarbon derivatives characterized by the presence a carbonyl group that react with two hydrocarbon backbones. Once again, the shorter-chain ketones are flammable liquids, and the longer-chain ketones are combustible solids. They include acetone (dimethyl ketone); 2,3-butanedione; cyclohexanone; methyl ethyl ketone (better known as MEK); methyl butyl ketone, methyl isoamyl ketone, methyl isobutyl ketone (MIBK); mesityl oxide (a ketone). and methyl propyl ketone. The proper chemical names for the ketones now all end in -one.

The **aldehydes** are a group of hydrocarbon derivatives characterized by the presence of the carbonyl group, as in the ketones, but with a hydrogen atom attached to the carbon rather than a second hydrocarbon radical. The aldehydes as a class have flash points similar to the ketones, if just slightly lower; and formaldehyde, the simplest, is a gas, although it is usually transported dissolved in water and called formalin. Common aldehydes that are flammable liquids include acetaldehyde, acrolein, butyraldehyde, dimethylpentaldehyde, paraldehyde, propionaldehyde, isobutyraldehyde, isopentaldehyde, methylpentaldehyde, 2-ethylbutyraldehyde, 2-methyl butyraldehyde, 2-methylpropanal, and valeraldehyde. The proper chemical names for the aldehydes now all end in -al.

Amines are hydrocarbon derivatives that have the general formula $R–NH_2$, but they may also be considered to be derivatives of ammonia (NH_3), with a hydrocarbon radical replacing one of the hydrogen atoms of the ammonia. Common amines that are flammable liquids include allyl amine, amyl amine, sec-amyl amine, butyl amine (and isomers), cyclohexylamine, ethyl amine, ethyl butyl amine, diethyl amine, triethyl amine, furfurylamine, hexylamine (and isomers), propyl amine, isopropyl amine, dipropyl amine, diisopropyl amine, butyl amine, and sec-butyl amine. The longer-chain amines are combustible liquids.

The **esters** are the flavors and fragrances of nature. Some are very disagreeable, while others are familiar and quite pleasant. All the fragrances and tastes of nature are due to esters. Common esters that are flammable liquids are methyl formate, methyl acetate, methyl acrylate, methyl propionate, methyl butyrate, ethyl formate, ethyl acetate,

and any isomers of these esters. The longer-chain esters are combustible liquids.

Combustible liquids

Combustible liquids are usually less hazardous than flammable liquids because of their higher flash points, but the spread should be significant, such as 10°F or 15°F. But once the flash point of a combustible liquid has been reached, it usually takes less energy to ignite its vapors than those of a flammable liquid, because combustible liquids usually have a lower ignition temperature, especially within an analogous series. There may be fewer vapors in the air, but by definition, the vapors of both liquids are present in sufficient quantities to burn. Remember that what always happens first when vapors that burn are ignited is an explosion. The fire usually follows.

Compressed gases

Gases are the least dense of the three states of matter (gas, liquid, and solid) and consequently gases have much more freedom to move around than matter in either of the other two states. Gases will always fill and take the shape of the containers in which they are placed. Their molecules will move randomly and endlessly in all directions within the container. It can be said that gases are somewhat related to liquids, because they are both fluids. Gases are defined as fluids with vapor pressures higher than 40 psia (pounds per square inch absolute) at 100°F; this definition includes the vapor pressure of air, 14.7 psi. This definition may seem to be unnecessary, but it becomes extremely important when the hazardous material with which we are confronted is in liquefied form. If the material is indeed a gas in its natural form (i.e., it is a gas at room temperature and normal atmospheric pressure), it must be realized that the gas (as a liquid) is actually confined in the container at a temperature above its boiling point. This means that if there were a sudden drop in pressure above the liquid, such as would occur if there were a leak in the container, the liquefied gas would attempt to convert to the gaseous state as rapidly as possible. This occurs because the gas would be following the natural law that a material can never reach a temperature higher than its boiling point, except when it is under pressure.

Compressed gases come in two states, pressurized and liquefied. Pressurized gases are those gases whose boiling points are either extremely low, around −150°F, or those gases that the seller does not wish to liquefy because the buyer is satisfied with the gas in its pressurized form (i.e., the gas in the container is still in gaseous form, except that it is now under a pressure higher than atmospheric pressure). Liquefied gases are those gases whose boiling points are higher, from −150°F to about 32°F. Gases are liquefied to save transportation space.

Because of safety concerns caused by increases in pressure, containers for compressed gases come with various safety relief devices. Most such containers have a spring-loaded valve, which is preset to activate at some elevated pressure, but one which is still well below the design strength of its container. For example, a container may be built to withstand pressures of 1,000 psi, but the spring-loaded valve will be set to activate at 250 psi. As the spring-loaded valve opens, the gas within the container is vented until its pressure drops below 250 psi, at which point the valve reseats itself, and the container is sealed once again. A spring-loaded valve is the only safety relief device that can reset itself after the gas has been evacuated, to keep the container from completely emptying itself. This is not the case with the other two common safety relief devices, the frangible disc (frangible means capable of bursting) and the fusible plug. The frangible disc will rupture at a predetermined pressure and will allow the container to empty itself completely. The fusible plug is designed to melt at a predetermined temperature, and when it does so, the container will also empty itself. The only pressurized container not to have a pressure-relief device is one designed to hold a poison gas. Here, the danger is greater from venting the tank than from a violent tank failure.

There is a correlation between the temperature, pressure, and volume of a gas in its container. When temperature is increased on a gas in a container, the pressure of the gas increases, and pressure decreases when the gas is cooled. Conversely, the gas heats up when pressure on the gas is increased, and the gas cools down when pressure drops. If the volume of the container decreases (as in a piston moving in a cylinder, the pressure (and temperature) of the gas increases, and if the volume of the container increases, the pressure (and temperature) of the gas decreases.

There are three gas laws that can be used to calculate changes in any of the variables when certain information is known. Boyle's, Charles's, and Gay-Lussac's laws are used, and have been brought together in the combined gas law. The interested student may consult any chemistry or physics textbook to better understand these relationships.

Compressed gases, like other classes of hazardous materials, have certain hazards associated with them. No particular gas possesses all the hazards listed, but many of them have more than one hazard. It is extremely important to realize that there is a list of hazards with which every emergency responder must become familiar. These are the hazards associated with compressed gases:

- They are all under pressure.

- They may be flammable (e.g., natural gas, the first four alkanes, LP gas, the first three alkenes, acetylene, hydrogen, ethylene oxide, carbon monoxide, hydrogen sulfide, hydrogen cyanide, cyanogen, formaldehyde, vinyl chloride, ammonia)

- They may be unstable (e.g., acetylene, ethene, propene, butene, vinyl chloride)

- They may be toxic (e.g., carbon monoxide, hydrogen sulfide, fluorine, chlorine, hydrogen cyanide, cyanogen, vinyl chloride, sulfur dioxide, the nitrogen oxides, ozone)

- They may be corrosive (e.g., fluorine, chlorine, hydrogen fluoride, hydrogen chloride, hydrogen bromide, hydrogen iodide)

- They may be oxidizing agents (e.g., oxygen, fluorine, chlorine, the nitrogen oxides, ozone)

- They may be subject to BLEVE-Boiling Liquid Expanding Vapor Explosion (any liquefied flammable gas)

Note: Ammonia is listed by the U.S. Department of Transportation (DOT) as a nonflammable gas because it has arbitrarily set specific lower explosive (flammable) limits for flammable gases. Ammonia has a lower explosive limit just outside the range set by DOT, but it will and does explode and burn!

Other gases have no particular hazard other than they might be asphyxiants when they rise above a certain percentage in air, and they are shipped and stored as pressurized gases. Examples are nitrogen, carbon dioxide, helium, neon, argon, krypton, and refrigerant gases. Some of these plus a few others may be shipped as cryogenic gases. (See section on cryogenic gases below.)

Flammable and combustible solids

This hazard class consists of ordinary Class A combustibles (not including plastics, which are Class A combustibles but are discussed separately), plus elements that burn. This includes wood and wood products; paper and paper products; carbon; sulfur; phosphorus; and almost all metals, especially lithium, sodium, potassium, aluminum, magnesium, zinc, and iron. All solid hydrocarbons, almost all solid hydrocarbon derivatives (especially nitrates such as cellulose nitrate), and, indeed, all organic material will burn. Most of the danger from these materials comes from the combustion products, including carbon, carbon monoxide, carbon dioxide, and water. Incomplete combustion products include many toxic compounds, most of which have not been identified because they, too, are combustible.

Cryogenic gases

The only difference between a normal liquefied gas and a cryogenic gas is the boiling point. The definition the term cryogenic refers to the behavior of materials from −150°F down to absolute zero (−459.67°F or −273.15°C). The most common cryogenic gases are oxygen, nitrogen, hydrogen, natural gas, helium, neon, argon, krypton, xenon, and fluorine, all of which have boiling points below −150°F. Cryogenic gases are transported and stored in what amounts to giant insulated bottles. When the container is small, it is called a Dewar's flask. These do not resemble other containers for pressurized liquids or gases, because the vapor pressure above a cryogenic liquid is very low, often less than a few pounds per square inch! The containers are really one container within another, using an air space between the container walls as insulation. Cryogenic liquids are basically self-insulating. Heat energy is required whenever evaporation occurs (the latent heat of vaporization of a liquid). This means that for the cryogenic gas to convert from a liquid to a gas, heat must be absorbed by the liquid so that some of it may evaporate. The heat must be absorbed from somewhere, and since the container is insulated and the cryogenic liquid is very cold, there is very little heat energy to supply for the evaporation process. With so little energy available, evaporation occurs very slowly. Even when a container of a cryogenic liquid is opened to the atmosphere, evaporation occurs so slowly that it is difficult to realize that the liquid is boiling. However, when there is a failure of the container or other event that causes a release of the cryogenic liquid to the environment, the boiling and release of gas will be much faster. Safety relief devices are used in all cryogenic liquid containers, with one in the internal container to vent to the atmosphere what little gas has evaporated, and another in the outer container to vent any gas that may have accumulated between the walls of the inner and outer containers. If this second safety relief device vents,

it is usually because there is a leak in the inner container. An alarm is normally connected to the outer safety relief device to alert workers to the leak, which could be extremely hazardous. The hazards, of course, depend on the nature of the gas. in any event, any gas leaking will be extremely cold.

Hazards of cryogenics include:

- The extreme cold
- The high liquid-to-vapor ratio
- The hazard of the particular gas itself

The hazard of the particular gas is whatever hazards the gas may have before it was liquefied. Liquid oxygen is a powerful oxidizer, as is liquid fluorine. Liquid fluorine is also corrosive. Liquid natural gas (LNG) and liquid hydrogen are flammable, and liquid nitrogen, LNG, hydrogen, helium, neon, argon, krypton, and xenon are all simple asphyxiants.

Oxidizing agents

An oxidizing agent, or oxidizer, is defined as a substance that contains oxygen and gives it up readily, or will otherwise support combustion. This definition includes fluorine, chlorine, bromine, and iodine, which will all support combustion. There are some substances that contain oxygen, like carbon dioxide and silicon dioxide (sand) that are not oxidizing agents because they will not give up the oxygen readily or easily. Actually, these materials are good fire-extinguishing agents. The danger of oxidizers, of course, is that when they mix with ordinary combustibles, they can produce raging fires with an almost unlimited supply of oxygen or other oxidizer. Some combustibles may become explosive in this situation. There are both organic and inorganic oxidizing agents. As a rule, the inorganic (or ionic) oxidizers will not burn (ammonium nitrate and other ammonium oxysalts are glaring exceptions), whereas the organic oxidizers will not only burn, many of them will self-destruct or explode. The organic oxidizers will be discussed under unstable materials. Oxygen from the atmosphere is the most common oxidizer on Earth. Next comes oxygen provided by substances (oxidizers) that contain it chemically, but give it up readily. We are provided a hint in the chemical name of such substances. The chemical name of almost every inorganic oxidizer ends in either -ate or -ite. This means oxygen is present in the compound. The only hitch is that not every chemical ending in -ate or -ite is an oxidizing agent. Some of these compounds hold onto their oxygen so tightly that they never give it up easily.

All ammonium compounds with oxyanions are powerful oxidizers that burn and, at some point, may explode. Ammonium nitrate mixed with an organic material is a powerful blasting agent and is a favorite weapon of terrorists.

The halogens, fluorine, chlorine, bromine, and iodine are all oxidizing agents, with fluorine being the most powerful of all. These oxidizers contain no oxygen, but they are very efficient in supporting combustion.

Corrosives

A corrosive may be defined as a chemical agent that reacts with the surface of a material, whether living tissue or not, causing it to deteriorate or irreversibly wear away at the site of contact. Corrosives may have a powerful destructive rate on live tissue, metals, and other materials. There are two major classes of corrosives, plus some corrosive materials that are outside these two classes. The two classes are acids and bases. An acid is defined as a chemical compound that contains one or more hydrogen ion that will liberate hydrogen gas on contact with certain metals, has a pH of less than 7.0, and may be very active chemically. A base is defined as a chemical compound that contains the hydroxide ion, is the chemical opposite of an inorganic acid, is active chemically, and has a pH greater than 7.0. When an acid reacts with a hydroxide, it forms a salt and water (the neutralization reaction). Some of the bases are as corrosive as strong acids. A base is the chemical opposite of an acid, and is often called an alkali.

Inorganic acids are usually stronger than organic acids. That is, they will ionize to a greater degree than organic acids. The more concentrated an acid, the more dangerous it is. Concentration of an acid is determined by how much acid is dissolved in water. The acids present one of the longest lists of hazards of any class of hazardous materials. Again, just as in lists of the hazards of other hazardous materials, it is extremely rare, if not impossible, for a member of that hazard class to possess all the hazards attributed to the class. For example, the hazards of inorganic and organic acids are different. The hazards of the acids (including both inorganic and organic acids) include the following:

- Corrosive (e.g., hydrofluoric, hydrochloric, hydrobromic, hydriodic, sulfuric, and nitric acid)
- Explosive (e.g., perchloric and picric acid)
- Polymerizable (e.g., acrylic and methacrylic acid)
- Oxidizers (e.g., nitric, nitrous, perchloric, chloric, chlorous, hypochlorous, chlorosulfonic, chromic, perchromic, and peracetic acid)
- Water-reactive (e.g., sulfuric acid)
- Toxic (e.g., hydrocyanic, arsenic, chlorosulfonic, carbolic, and formic acid)
- Combustible (e.g., all the organic acids)
- Very reactive (e.g., the inorganic acids)
- Unstable (e.g., perchloric, persulfonic, and peracetic acid)

Bases include the hydroxides of all the metals plus ammonium. Sodium and potassium hydroxides are the two most common alkalis and probably the strongest bases. The hydroxides of the alkali metals are by far the most corrosive, followed by the alkaline earth metal hydroxides, then aluminum hydroxide. Concentrated ammonium hydroxide is very corrosive.

Organic peroxides

The other unstable hazardous materials are the organic peroxides. There are covalent compounds containing the peroxide radical with hydrogen or hydrocarbon derivatives attached to the oxygen atoms. Although these peroxides are oxidizing agents, they are more dangerous than the inorganic peroxides listed with other oxidizers. Organic peroxides are all synthetic and are used as initiators or catalysts in polymerization reactions. They are so reactive that an inhibitor must be added to the reactor to keep the reaction from starting prematurely. An inhibitor is defined as a substance used to slow down or prevent a chemical reaction. An initiator is defined as a chemical added to a monomer that will overcome the stabilization action of the inhibitor to allow the polymerization reaction to begin. The initiator may also be called a starter. The organic peroxides are generally liquids or white powders, depending on their individual chemistry. The peroxide functional group between two hydrocarbon radicals is extremely unstable. The slightest amount of energy input to the molecule may be enough to cause a violent molecular decomposition with the release of a tremendous amount of energy. The organic peroxide molecules are fairly stable at some low temperature, and they are usually kept at that temperature. Organic peroxides must be stored properly. The more dangerous of them must be kept cool, inside a special isolated storage building with no windows, no electrical switches on the inside, and limited access to personnel. There are newer organic peroxides on the market today, specially stabilized and formulated so that they need not be so restricted; but as a rule, they are not as effective and efficient as the more hazardous organic peroxides.

The organic peroxides have many other hazards including:

- They are unstable.

- They are flammable (in the sense that they may self-ignite).

- They are highly reactive.

- They may be explosive (especially in a fire).

- They are corrosive.

- They may be toxic.

- They are all oxidizers.

TOXIC MATERIALS

Toxicity may be defined as the ability of a chemical substance to produce injury once it reaches a susceptible site in or near the body. Alternatively, it is the harmful effects of a chemical on some biologic mechanism. A poison is defined as any substance that causes injury, illness, or death to living tissue by chemical activity. Certainly, toxicity is one of the most controversial hazards. There is no question that carbon monoxide is highly toxic, as well as many other chemical substances. However, it takes a certain level of CO to cause harm, and that harm may be reversed. There are also many foods and other consumer products that have harmful effects on humans in one manner or another. The point is that there are many factors of toxicity that can determine the degree of harm that can be caused to humans. Dosage may be the most important determinant in exposures to toxic materials. It is defined as the actual quantity of a chemical administered to an organism or to which the organism is exposed. Exposure is an important factor. Whether the exposure is acute or chronic is a determinant. Many materials have a cumulative effect on humans, and the number of exposures may be critical. The entry route of the toxic material into the body is very important. Normally, toxic materials enter the body in one of four ways: inhalation, ingestion, absorption through the skin, and entry through a wound. There are other methods of entry, but these four are the most common. The method of entry may determine how rapidly the toxic nature of the poison will produce its effects. The body chemistry of the victim is a determining factor in how certain materials will affect that body as well. Many attempts have been made to determine the actual amount of toxic material that will be harmful to humans. Tests are often incorporated into laws, standards, and regulations meant to protect workers who are regularly exposed to materials thought to be toxic. Some of those measurements are:

LD_{50} (lethal dose) is defined as the amount of material, usually expressed as milligrams per kilogram of body weight (mg/kg), that will cause death to half the test animals to which the material was administered (usually fed) during the test period, usually 14 days.

LC_{50} (lethal concentration) is defined as the amount of gases, vapors, fumes, dusts, and other particulates that will cause death to half the test animals exposed to the material within a specified time.

TLV-TWA (threshold limit value–time weighted average) is defined as the weighted average amount of a material to which a person may be exposed over a normal 8-hour workday or 40-hour work week without harm.

STEL (short-term exposure limit) is defined as usually a 15 minute time-weighted average exposure which should not be exceeded at any time during a workday, even if the 8 hour time-weighted average is within the TLV-TWA.

IDLH (immediately dangerous to life and health) is defined as the amount of material, usually expressed as parts per million (ppm) of air or other material or milligrams per cubic meter of air (mg/m^3), that will pose an immediate life- or health-threatening situation to the exposed person.

RADIOACTIVITY

Radioactivity is defined as the spontaneous emission of radiation from the nucleus of an unstable atom. It is the least understood of all the hazards that face firefighters who respond to hazardous materials incidents. This is so because the hazard is invisible and appears to be very difficult to understand. However, it is not necessary to understand all the mechanism of radioactivity to be able to protect yourself at an incident involving radioactive materials.

Accidents involving radioactive materials are very rare, and the safeguards built in to prevent the escape of radiation are very effective. Containers used to ship radioactive materials are built to prevent leaks in the worst of accidents, and nuclear power plants have many redundant backup safety systems built into their operations. The quantities of radioactive materials found at hospitals, universities, research facilities, and industrial sites are usually very small. But the specter of Murphy and his infamous law hangs over first responders, and the fact is that an accident will occur, some time, somewhere, that will require the hazardous materials response team.

There are two types of radiation, ionizing and nonionizing. Nonionizing radiation takes the form of waves of energy such as radiant heat, radio waves, and light. The amount of energy in these waves is small when compared to ionizing radiation. Ionizing radiation involves particles traveling in wavelike motions. The collision of these particles with atoms in the victim's body is what causes the damage, so methods of protection must be sought. These particles originate in the nucleus of an unstable atom, and the most important are alpha particles, beta particles, gamma rays (also particles), and neutrons.

The alpha particle is the largest of the four types of radiation particles. It is composed of two neutrons and two protons and therefore has an atomic weight of 4 and an electrical charge of + 2. It is identical to the nucleus of the helium atom (indeed, if helium's two electrons are removed from the atom, what will be left is an alpha particle). When it is emitted from the nucleus of one of the large, heavy atoms, it seeks out two electrons wherever it can find them to make itself electrically neutral. When it gains these electrons, it indeed becomes an atom of helium. Alpha particles are emitted by nuclei that are trying to reach stability by transmutation to another element. The alpha particle is by far the largest of the particles of radiation. Because of its size, it has the potential to do the most damage by ionization, simply because the probability of collision of alpha particles with an electron is higher than of the other forms of radiation. However, also because of its size and the probability of collision with the electrons of other atoms, particularly those of air, the alpha particle will not travel very far from its source. The distance is usually 3 to 4 inches. Normally, an alpha particle will not penetrate the epidermis, the layer of dead skin surrounding our bodies, and it can be stopped by a thin sheet of paper. Normal turnout gear will stop alpha particles. Care must be taken not to inhale or ingest an alpha particle emitter, since it will never be more than a few inches from some

vital organ, and there is no protection against the alpha particle inside the body.

The beta particle is a particle that has the same mass as an electron and will have in some cases the same charge (−1) and in other cases the opposite charge (+1). The beta particle is quite small when compared to the alpha particle, and it can travel much farther than the alpha particle before it collides with an electron and forms an ion pair. The beta particle can travel up to 100 feet in air and will penetrate the skin layer, but it can be stopped by a thin piece of metal or by an inch or two of wood. It was originally felt that full turnout gear would stop beta particles, but recent tests indicate that normal turnout gear may not be effective at all in stopping beta particles. To be on the conservative side, you must assume that your full turnout gear will not protect you from beta radiation. You may have to use special protective equipment.

The third method of radioactive decay (and the most dangerous) is gamma radiation. It may be easier to define gamma radiation as electromagnetic radiation of a considerably shorter wavelength and a considerably higher energy than light, heat, radio waves, and X-rays, which are also forms of electromagnetic radiation. The fission reaction (the "splitting" of nuclei) is a prime source of gamma radiation. The damage done by gamma radiation is ionization, just as in the case of alpha and beta emission. In this case, the gamma radiation possesses so much energy that it will "grab" an electron and leave an ion behind. Gamma radiation is considered by many to be pure energy and therefore can travel great distances, perhaps miles, at the speed of light (approximately 186,000 miles per second). Because of size and energy of gamma particles, gamma radiation is very deeply penetrating and will pass through any barrier that will routinely stop alpha and beta particles. This means that several feet of earth or concrete or several inches of lead will be needed to stop gamma radiation. The implication is that although your turnout gear may protect you from alpha particles, and you may wear special protective clothing to stop beta particles, there is nothing you can wear to protect you from gamma radiation that will still allow you to move about.

The fourth form of radiation is neutron radiation. The neutron is the neutral particle in the nucleus and is released by the explosion of a nuclear device. It is an extremely rare form of radiation, but it will be present in the event of discharge of atomic weapons. The neutron is a particle considerably larger than the beta particle but has only one-fourth the mass of the alpha particle.

It is rare, but is now a new threat because of the possibility of nuclear detonation by terrorists. It is very deeply penetrating and therefore highly dangerous, the type of radiation that will cause the most deaths among the survivors of the discharge of an atomic weapon. Lead shielding may be necessary for protection against neutron radiation.

Everyone recognizes the rather self-explanatory motto, "time, shielding, and distance" as the proper way to gain protection from radiation dangers. Time means simply to limit your exposure to radiation sources for as short a period of time as possible. Dosimeters worn on protective clothing and a timekeeper will indicate how long a safe exposure is. Shielding indicates the proper protective material to be kept between the radiation source and the responder, and distance indicates how much distance is needed from a particular radiation source.

There are several measurements of exposure and absorption of the different types of radiation. They include the rad, rem, roentgen, and curie. The rad (radiation absorbed dose) is the unit of radiation absorbed dose. The rem (radiation equivalent man) is a measure of radiation dose related to biological effect. The roentgen (R) is the unit of exposure from X or gamma rays, and the curie (Ci) is the unit of activity defined as the number of nuclei in a sample that disintegrates every second (3.7^{10}). It defines the rate of decay of a radioactive material.

EXPLOSIVES

An explosive is defined as any pure substance or mixture whose principal purpose is to explode. Not every material that explodes is an explosive. For example, natural gas, when it is in the flammable range and has accumulated, will explode upon ignition, as will the vapors of gasoline, but they are not classified as explosives. An explosion is defined as the phenomenon characterized by the instantaneous release of heat and pressure caused by the oxidation of fuels or the decomposition of molecules. Alternatively, it is the sudden and violent release of mechanical, chemical, or nuclear energy from a confined space.

There are six basic types of explosions. They are pressure-relief, rapid oxidation, molecular decomposition, runaway polymerization, nuclear fission, and nuclear fusion. Discussion will be limited to the first four types.

The pressure-relief explosion is really a mechanical operation, the result of the failure of a container due to the rise of pressure above the container's design strength. A typical pressure-relief explosion is the failure of a boiler or some other heating system caused by overheating and/or the failure of a pressure-relief device to operate properly. It can also be caused by the weakening of the container fabric by corrosion, metal fatigue caused by excessive heat, or physical damage. Also, the first stage of a BLEVE is a pressure-relief explosion, caused by the concurrent rise in internal pressure, caused in turn by applied heat energy from another source and the subsequent weakening of the metal container by an impinging flame.

The rapid oxidation explosion is a chemical reaction whereby a flammable gas or the vapors of a flammable liquid are within the explosive range, and a suitable ignition source ignites the vapors or gas. It is also the chemical reaction in which solid or liquid fuels are intimately mixed with an oxidizing agent (as in the case of gunpowder or blasting powder), and the mixture is shocked or otherwise detonated.

The molecular decomposition explosion is an instantaneous release of heat and pressure brought about by the simultaneous disintegration of covalent bonds, such as in the explosion of nitroglycerine. A molecular decomposition explosion of a substance that is not an explosive is characterized by the breakdown of the triple bond in the acetylene molecule that occurs when acetylene is shocked.

In the runaway polymerization explosion, there is an instantaneous release of heat energy from the chemical reaction of monomers with themselves. The breaking of covalent bonds (basically, the double bonds in the monomer) starts a type of chain reaction that, if not controlled, will explosively destroy whatever container that holds the reaction.

The pressure-relief explosion is not a chemical reaction, and the runaway polymerization explosion is a very specialized explosion involving only very special small molecules called monomers. The two nuclear explosions, fission and fusion are also not chemical reactions, so the explosives explored will be those nonmilitary explosives that operate by rapid oxidation or molecular decomposition.

Explosives are divided into high and low explosives, based on the speed of reaction. Although there is no fine dividing line between the two classes, the high explosives generally detonate through a supersonic propagation of the shock wave through the material, while low explosives are characterized by deflagration, a subsonic propagation of the shock wave through the material.

High-order explosives (or simply, high explosives) are divided into three subclasses: main charges, primary explosives and secondary explosives. A main charge is a high explosive that is relatively insensitive to heat or shock. These explosives usually are the most powerful of the explosives in their brisance; that is, the sharp, shattering effect an explosive has on its surroundings. They are so insensitive to heat and shock, however, that it usually takes another high explosive to cause them to detonate. Examples of high explosives that are used as the main charge include TNT (2,4,6-trinitrotoluene), nitroglycerine, dynamite (nitroglycerine absorbed onto an inert material), picric acid, ammonium picrate, Composition B, HBX-1, EDNA (ethylenedinitramine), gelatin dynamite, and the "plastic" explosives C-3 and C-4.

Primary explosives are high-order explosives that are extremely sensitive to heat and shock. For this reason they are among the most dangerous of the high explosives. Because of this sensitivity they are absolutely useless as the main charge, but they are useful for initiating the detonation of high explosives used as the main charge. Furthermore, they do not have the brisance, or explosive power, that the high explosives do, but they are powerful enough to detonate the main charge. They are also known as initiators or detonators. The most common examples of this group are mercury fulminate, silver fulminate, and lead azide.

Secondary explosives are high-order explosives that are less sensitive to heat and shock than the primary explosives. Secondary explosives are used in those situations where the sensitivity of the main-charge explosives is such that the power of the primary explosive is not enough to cause detonation of the main charge. The secondary explosive can be detonated by the primary, and in turn it will detonate the main charge. In boosting the relatively weak power of the primary explosive, the secondary explosives have become known as "booster" explosives. When the primary explosive, the secondary explosive, and the main charge are arranged so that the primary explosive detonates the secondary explosive, which in turn detonates the main charge (all of which occurs, of course, in milliseconds), this sequence of events is known as an explosive train. Some examples of secondary explosives are RDX (cyclotrimethylenenitramine), tetryl (trinitrophenylmethylnitramine), lead styphnate (lead trinitroresourcinate), and PETN (pentaerythritoltetranitrate).

Low-order explosives (or low explosives) are those explosives that deflagrate rather than detonate, unless they are confined, which they always are. The low explosives are also called propellants, since that is, or originally was, their main function. There are only three principal explosives in this class: black powder, smokeless powder, and cordite. Of these three, black powder is very dangerous because of its sensitivity to heat, shock, or energy of any kind. Black powder was once the explosive of choice for terrorists. It is so sensitive that they have switched to a blasting agent such as ANFO (ammonium nitrate-fuel oil).

It is difficult to keep all the newer explosives straight, since they all have brand names. There are many other types of commercial explosives, each developed and designed to do a specific job. These explosives are usually organic compounds containing the azide, bromate, chlorate, chlorite, fulminate, iodate, nitrate, nitrite, perchlorate, or picrate functional groups. Beware of any compound that has an organic name (methyl, ethyl, phenyl, and so on, somewhere in the name) and one of the above-mentioned functional groups in it, since it must be treated as an explosive until it can be proved otherwise. Another hint is the presence of the nitro group in the name. Not all the organic nitro compounds are explosives, but most are!

A blasting agent is a material designed to explode but not classified as an explosive. It is a mixture of materials intended to have enough power to shatter or loosen something, usually in a construction project, but it is so insensitive to shock that even a powerful blasting cap (which is an explosive) will not make it explode. Ordinarily, a blasting agent requires a high explosive like dynamite to detonate it. To be classified as a blasting agent, the mixture must not contain any material that can be classified as an explosive. Terrorists have learned to make blasting agents and how to use other explosives. We have been attacked by them twice in this country using ANFO, and we must be prepared for more attacks using blasting agents and explosives.

SYSTEMIC POISONS

A material is defined as a systemic poison if it interferes with any vital bodily processes. Systemic poisons include those that interfere with the function of the liver and kidneys, bone marrow, muscles, and vital nerve impulses. Arsenic and arsenic compounds are the most common systemic poisons that interfere with the function of the liver and kidneys. The heavy metals, including lead, cadmium, and mercury attack the liver and kidneys, as well as the halogenated hydrocarbons. The most common poison to attack bone marrow is benzene, which is the most dangerous of a group of aromatic hydrocarbons including toluene, the xylenes, and naphthalene. By far, the most common poison that attacks muscle groups is strychnine, which in severe poisoning cases can cause the victim's muscles to contract with such power that the bones will break. The largest group of poisons that interfere with vital nerve impulses are the organic phosphates, materials commonly used in pesticides. It is not possible to list all the nerve impulses with which they interfere, but in high enough dosages the resulting death will be quite unpleasant. A new group of complicated organic chemical compounds that include the phosphate radical has been developed specifically for the purpose of poisoning large groups of people. These are commonly known as nerve agents, and include some now well-known names as sarin, tabun, soman, and VX. Other specific poisons that target specific nerve impulses are carbon disulfide, which first attacks the respiratory system before moving to other nerves, and methyl alcohol (methanol), which attacks the optic nerve first before other targets. There are many other poisons that end up shutting down the central nervous system (CNS), causing great damage and eventually death. Poison A is a DOT classification that includes extremely dangerous poisons, substances that are poisonous gases or liquids such that a very small amount of the gas, or vapor of the liquid, mixed with air is dangerous to life. Poison B is a DOT classification that includes less dangerous poisons, substances, liquids, and solids (including pastes and semisolids), other than Poison A or irritating materials, which are known to be so toxic to humans as to afford a hazard to health during transportation, or which in the absence of adequate data on human toxicity are presumed to be toxic to humans.

WATER-REACTIVE MATERIALS

Firefighters, in the vast majority of responses, fight fires with water. There are, however, some hazardous materials that will react in a highly detrimental manner when they come into contact with water. In some cases, these materials will break into flames, produce toxic or flammable gases or vapors, involve tremendous amounts of heat, or produce some other adverse effect. As a class, they are known as water-reactive materials or, more simply, water-reactives. DOT classified a few of the materials below as flammable solids, even if the materials themselves do not burn. Their reasoning is that when such a material comes into contact with water, it will liberate a flammable gas that will usually be ignited by the heat also liberated by the reaction, or simply because extremely flammable gases are liberated. The "flammable solid" designation is usually accompanied by a "dangerous when wet" placard or label.

The alkali metals, the elements of Group IA on the periodic table of the elements, are all water-reactive materials. These elements are so reactive that they will seek out the oxygen in the water molecule and literally rip it from its chemical bonding with hydrogen, no easy feat. The action is exothermic, producing enough heat energy to ignite the hydrogen released in the same reaction. The reaction is so violent that, if the piece of metal is very large (about the size of a marble or larger), the metal will burst apart into smaller pieces, scattering these particles about and causing many more reactions wherever they come into contact with moisture. As the pieces of metal shrink in size during the reaction, they assume the shape of spheres, like tiny ball bearings, rolling around the surface of the water (lithium, sodium, and potassium all have specific gravities of less than 1.0), spitting and cracking as they react with the water. The flames of hydrogen will be very hot, and almost invisible in daylight. The corresponding metal hydroxide is formed in the remaining water, and this solution can be corrosive if the hydroxide concentration is very high.

The alkaline earth metals, the elements of Group IIA on the periodic table are less reactive than the alkali metals, but the pure metals will react with water. Luckily, here are also very few uses for the pure metals (with the exception of magnesium), so you stand little chance of encountering them.

The hydrides are a group of inorganic, water-reactive, binary ionic compounds that are made up of a metal and hydrogen. They are all irritants, toxic, and produce highly flammable reactions. Whenever they come into contact with water, they liberate hydrogen and the corresponding hydroxide. The explosiveness and flammability of hydrogen are well known, and the hydroxides of the alkali metals are extremely caustic (corrosive). Common among them are aluminum hydride, antimony hydride, arsenic hydride, barium hydride, beryllium hydride, calcium hydride, copper hydride, lead hydride, lithium hydride, magnesium hydride, sodium hydride, potassium hydride, and the boron hydrides.

The carbides are a family of binary compounds that contain carbon and another element. They are not all water-reactive, but each should be treated as such until you know that it isn't. Probably the most familiar of all the carbides is calcium carbide. It is a grayish black solid and has a faint odor of garlic. When it comes into contact with water, it liberates the highly flammable and unstable gas, acetylene, and a caustic material, calcium hydroxide. The water-reactive carbides include aluminum carbide, beryllium carbide, sodium carbide, and magnesium carbide.

The nitrides are binary compounds containing the nitride ion (N^{-3}). They are different from the azides, which contain the azide ion (N^{-1}). The nitrides and the azides are both hazardous materials, with the nitrides classified as water-reactive materials, whereas the azides are explosives. The water-reactive nitrides include aluminum nitride, beryllium nitride, lithium nitride, magnesium nitride, potassium nitride, and sodium nitride.

The phosphides are ionic compounds of a metal and phosphorus, with the phosphide ion being P^{-3}. When the phosphides react with water, they form the flammable and toxic gas phosphine, and the corresponding hydroxide. Water-reactive phosphides include aluminum phosphide, boron phosphide, calcium phosphide, and sodium phosphide.

There are many inorganic chlorides that are stable in contact with water, such as the chlorides of the alkali metals and the alkaline earth metals. Many of the others are water-reactive, including anhydrous (meaning no water present) aluminum chloride, antimony pentachloride, boron trichloride, stannic chloride, titanium tetrachloride, phosphorus oxychloride, phosphorus pentachloride, phosphorus trichloride, silicon tetrachloride, sulfur monochloride, sulfuryl chloride, thionyl chloride, and titanium dichloride.

The inorganic peroxides will react with water either to liberate hydrogen peroxide and the corresponding inorganic hydroxide, or it will liberate the hydroxide ion plus oxygen, depending on the temperature of the water. The peroxides are liberated with cold water; oxygen, with warm water. The peroxides of the alkaline earth metals may also have the same reactions. Remember that the principal hazard of the inorganic peroxides is their oxidizing power, and that water reactivity is secondary. The most common are sodium peroxide, lithium peroxide, and potassium peroxide.

PLASTICS/POLYMERS

Plastics are a part of a group of materials that belong to a much larger class of materials called polymers. A polymer is large molecule made up of thousands of tiny molecules that have the unique property of being able to react with themselves. The process by which this takes place is a specialized chemical reaction known as polymerization. The specialized tiny molecules (relatively speaking, since all molecules are tiny) that have this unique property of combining with themselves in the polymerization reaction are called monomers. Both words, monomer and polymer, come from Greek roots: mono means one, poly means many, and mer means part. Plastics are not hazardous materials when used as they are intended. However, hazardous materials are used in the polymerization process, namely, monomers and organic peroxides. These will be discussed in the unstable materials section. Plastics may be hazardous when they burn, producing products of combustion that may be irritating or toxic, or both.

UNSTABLE MATERIALS

Unstable materials may be defined as substances that decompose spontaneously, polymerize, or otherwise self-react in a hazardous manner. There is no specific DOT classification corresponding to the term unstable materials, but every hazardous material that will be discussed in this chapter is listed by DOT elsewhere, under a specific hazard. These substances are hazardous materials that are so dangerous that their classification in another area is insufficient to call their most dangerous hazards to your attention. These materials really fall into two general types of products, organic peroxides and monomers. The reason monomers are included as unstable materials is because of their ability to enter into a specialized chemical reaction called polymerization. During this polymerization reaction, which is intended to be carried out safely in a giant, sealed, jacketed vat called a reactor, large amounts of heat are generated (the reaction is exothermic as it gives of heat). This heat and the resulting rise in pressure caused by it are controlled in the reactor. If this reaction should take place, the evolved heat and pressure could result in a very dangerous explosion. The instability of monomers is due to the very property that allows them to polymerize: the double bond between carbon atoms. This is an extremely reactive site on the hydrocarbon chain that is used to produce the kind of chemical reaction necessary to make the desired product. Once this double bond is broken, it liberates enough energy to cause a chain reaction (not of the radioactive type) that causes the rapid release of energy.

The monomers have more than one hazard, including:

- They are unstable.
- They may be toxic.
- They may be corrosive.
- They may be under pressure.
- They are highly reactive.
- They are flammable.
- They may be subject to BLEVE.

IRRITANTS

The irritants are further broken down into water-soluble irritants such as halogen acid gases: hydrogen chloride (HCl), hydrogen fluoride (HF), hydrogen bromide (HBr), and hydrogen iodide (HI), as well as sulfur dioxide (SO_2) and ammonia (NH_3); moderately soluble irritants including the halogens (fluorine, chlorine, the fumes of bromine, and the vapors of iodine), ozone (O_3), phosphorus trichloride (PCl_3), and phosphorus penta-chloride (PCl_5); and the slightly soluble irritants, which include the nitrogen oxides: nitric oxide (NO), nitrogen dioxide (NO_2), nitrogen trioxide or nitrogen sesqui-oxide (N_2O_3), dinitrogen tetroxide or nitrogen peroxide (N_2O_4), dinitrogen pentoxide (N_2O_5), and trinitrogen tetroxide (N_3O_4).

ASPHYXIANTS

Asphyxiants are broken down into simple asphyxiants, blood asphyxiants, and tissue asphyxiants. A simple asphyxiant is a substance that will replace the oxygen in the air to a level below that required to sustain life. The most common simple asphyxiants are carbon dioxide and nitrogen. Hydrogen is a simple asphyxiant, but it rises in air and disperses so rapidly that it is not as dangerous as the other simple asphyxiants. The rest of the inert gases (neon, argon, and krypton) are simple asphyxiants. All of the saturated hydrocarbons are also simple asphyxi-

ants, which means that natural gas is not poisonous as it is portrayed in the movies.

Blood asphyxiants are materials that combine with red blood cells and render them incapable of combining with oxygen and thereby carrying the oxygen to the cells of the body. The most common of the blood asphyxiants is carbon monoxide, which is picked up by the blood cell at least 200 times more readily than oxygen. Two other common blood asphyxiants are aniline and nitrobenzene.

Tissue asphyxiants are materials that are picked up and carried by the red blood cells to the cells of the body, given up to those cells in exchange for the carbon dioxide held by those cells, thus poisoning them by making the body cells incapable of ever again accepting oxygen from the red blood cells. Hydrogen cyanide is the most common tissue asphyxiant.

RESPIRATORY PARALYZERS

Respiratory paralyzers are materials that will, upon entering the body, short-circuit the respiratory nervous system, causing respiratory distress. The most common respiratory paralyzer is hydrogen sulfide, a flammable, colorless gas with an overpowering smell of rotten eggs. Other common respiratory paralyzers are carbon disulfide, acetylene, and diethyl ether, commonly called just ether.

AIR-REACTIVE MATERIALS— PYROPHORICS

An air-reactive material is defined as a substance that will self-react and/or ignite when exposed to air. A pyrophoric material is defined as a material that ignites spontaneously upon exposure to air (or oxygen) at temperatures below 130°F (some sources quote 86°F). Thus, a pyrophoric material is an air-reactive material. Some of these materials may break into flame, some may decompose slightly less violently into some noxious components, and others may detonate. It is not possible to classify every material into what it does in contact with air, but here are pyrophoric materials you are likely to encounter: white or yellow phosphorous (red phosphorus is not pyrophoric); organometallic compounds such as tetraethyl lead, nickel carbonyl, and dimethyl arsine; the alkyl aluminums such as triethyl aluminum, trimethyl aluminum, trimethyl aluminum ethereate, chlorodiethyl aluminum, and tri(iso)butyl aluminum; dimethyl cadmium; and diethyl zinc.

CROSS-HAZARDS OF HAZARDOUS MATERIALS

It is extremely important to be aware that many hazardous materials have more than one hazard. That is, the principal hazard of a substance may be listed in a compilation of chemicals that have a particular hazard, such as toxicity, but miss a hazard that could kill a responder to the incident. An example is sulfuric acid, which is one of the largest volume chemicals manufactured in the world. Everyone recognizes that it is an extremely corrosive material, but you may miss the fact that it is a violent water-reactive material, and under certain conditions it may be an oxidizing agent. Toxic gases that are also flammable include carbon dioxide and hydrogen cyanide, two chemicals that are so toxic that their capability to burn (explode) may be overlooked.

Ammonia, a gas listed by DOT as nonflammable, will burn (explode) in concentrations where its toxicity would kill most responders without respiratory protection.

It is the responsibility of the leaders and trainers of hazardous materials response teams, and indeed any firefighter sent into a scenario where hazardous materials may be stored (or released), to train and educate these responders to all the possible dangers of the most common hazardous materials they may encounter. It is not enough to require minimum hazardous materials training for today's firefighters. The training, and especially the education, must be rigorous and thorough. This is absolutely necessary in today's environment where plans are being made for terrorist attacks disrupting our lives and even attempting to kill us with chemical weapons. Today's chemical warfare agents are yesterday's hazardous materials. The weapons used by the terrorists are chemicals that have been around longer than most firefighters and terror organizations are training with these materials in order to become more effective there we need to be well trained, aware and vigilant.

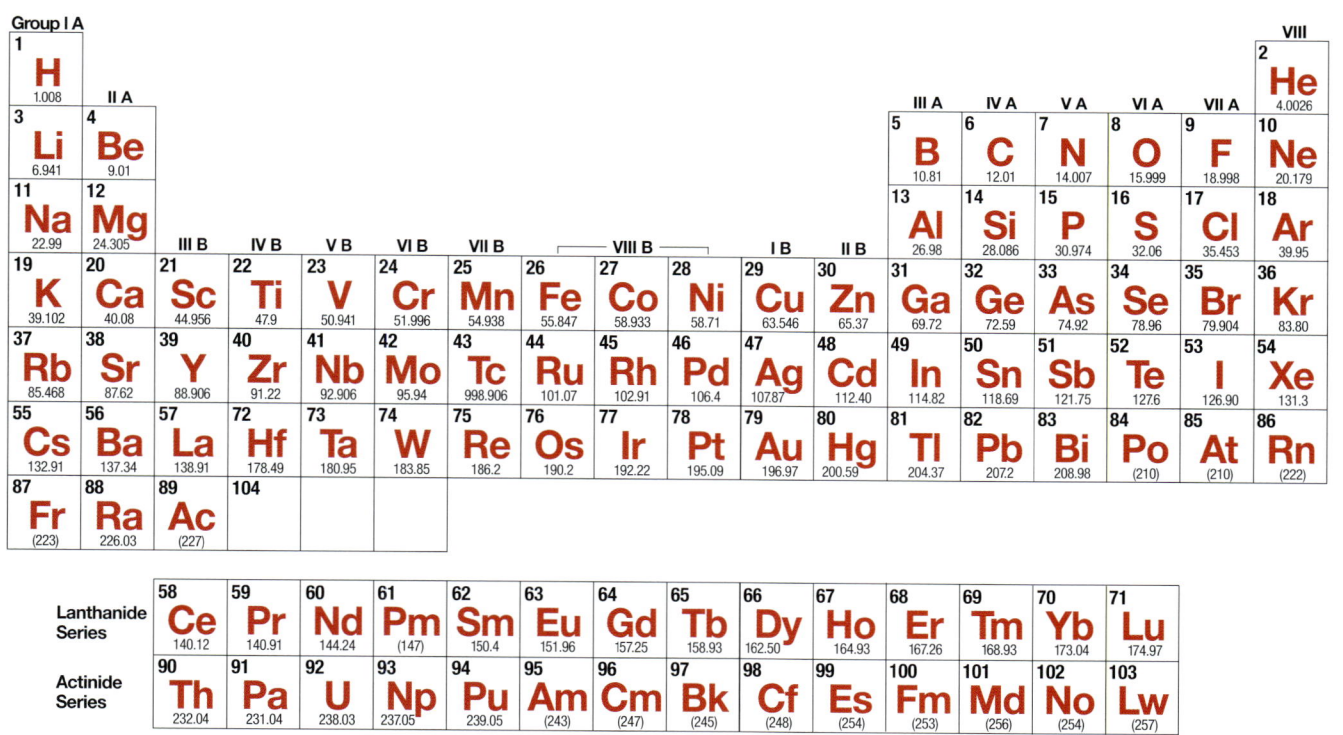

Fig. 6–1. The periodic table of elements

QUESTIONS

1. In chemistry, matter can be found in one of three states. Name them.

2. The Periodic Table of the Elements contains how many elements?

3. What is the definition of a chemical compound?

4. What is the one property common to all hydrocarbons?

5. Halogenated hydrocarbons have halogen atoms instead of what?

6. Ethers are considered especially dangerous because they form _____ when exposed to air.

7. Compressed gases come in two states: _____ and _____.

8. What is the one common hazard associated with all compressed gases?

9. Extremely cold materials are known as cryogenics. Their boiling points are below _____ °F.

10. One measurement of toxicity in materials is to determine the LD50 (lethal dose). Define how the LD50 is determined.

Skill Drills
Awareness Response to Hazardous Materials

These photos are for instruction only. Always follow your department's SOPs. When you perform these techniques at actual incidents, be sure to wear full protective gear and SCBA.

Recognition of Hazmat Markings

Step 1

- It is important to quickly recognize any potential hazmat situation early. Most materials will be marked with an appropriate marking system.

Step 2

- For transportation of hazardous materials, one type of system is the United Nations (UN) number. This is a four-digit number. It has a corresponding material that can be readily located using your ERG. North American (NA) numbers are identical to UN numbers with the exception that there are more NA numbers. NA numbers are a superset of UN numbers. These can also be found in the ERG.

Step 3

- Often the placards used during transportation will have a number at the bottom. These refer to the DOT hazard class. They go from 1 to 9, sometimes utilizing decimal points. These can also be found in the ERG.

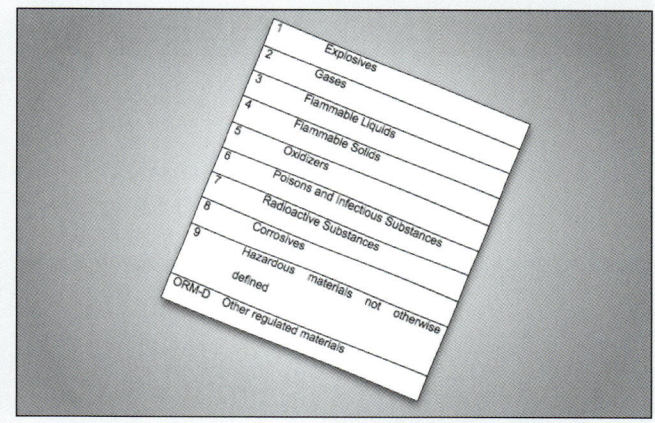

Step 4

- When dealing with fixed facilities, the NFPA 704 marking systems is used to denote the hazards present.

Step 5

- Take note of any pipeline markings denoting flow or contents.

Step 6

- Look for associated markings on packaging of materials. Often they will have UN numbers.

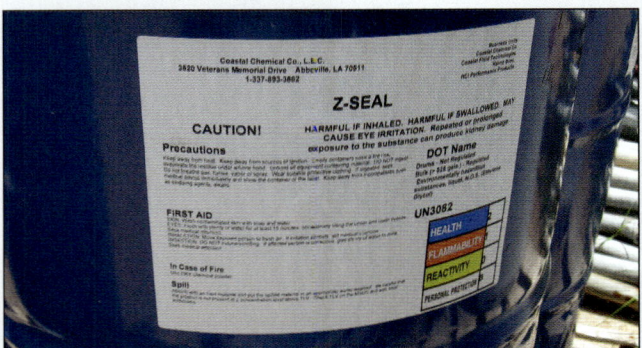

Step 7

- Military markings are used on military fixed facilities as well as on transport containers. It is important to be familiar with the various markings. In the event of an incident involving a military vehicle, notify the military immediately.

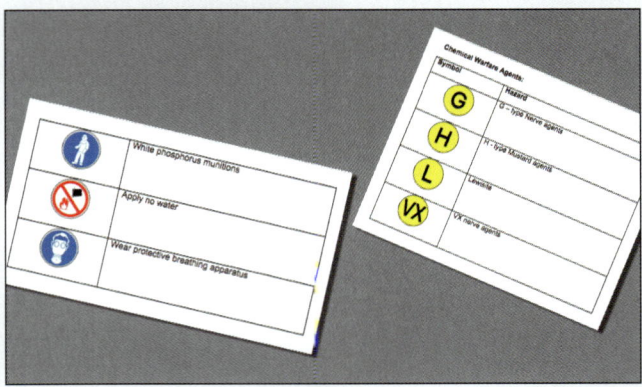

NFPA 704 Markings

Fixed facilities often have NFPA 704 markings denoting the hazards present on the premises. The NFPA 704 diamond is split up into four sections.

Step 1

- The blue section denotes the health hazard present. Zero denotes no hazard. The scale goes up to 4, which denotes a deadly health hazard.

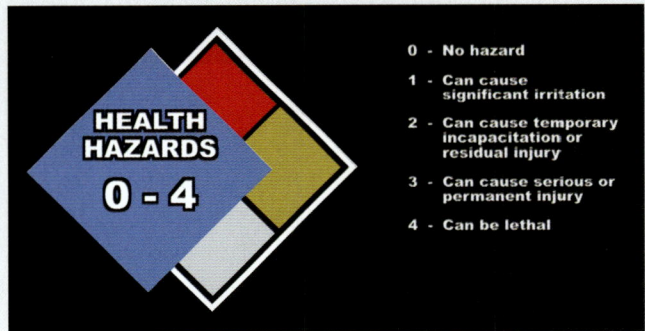

Step 2

- The red section indicates the fire hazard present. Zero denotes material that will not burn. The scale goes up to 4, which denotes material that will burn below 73°F.

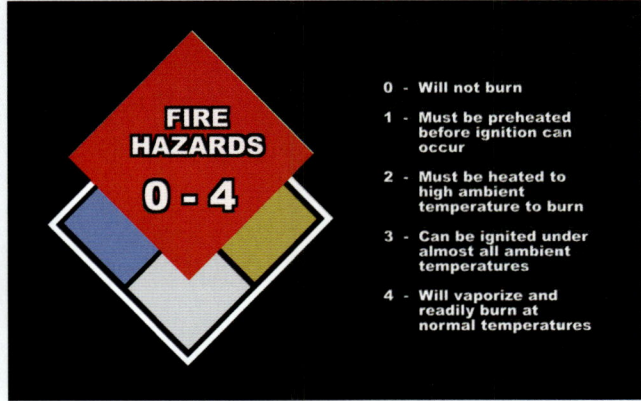

Step 3

- The yellow section denotes the materials reactivity. Zero denotes a stable material. The scale goes up to 4, which denotes a material that may explode.

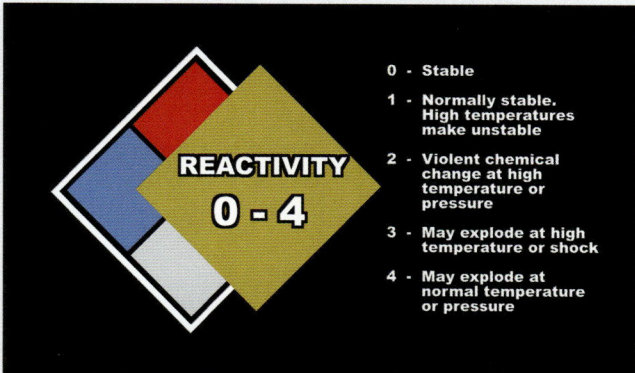

Step 4

- The white section is reserved to denote specific hazards such as acid, alkali, corrosives, oxidizers, radioactive material, as well as material that is water reactive.

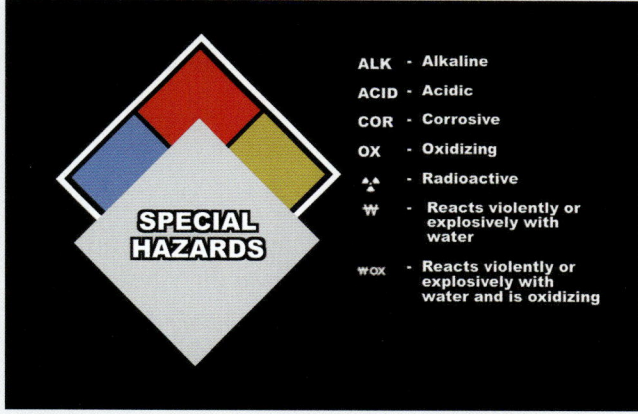

Analyzing a Pesticide Label

Step 1

- Pesticide labels contain the name of the pesticide and the following information:

Step 2

- A *signal word* is used to indicate how toxic the materials are. They are broken into the following categories: danger/poison is highly toxic, warning is moderately toxic, and caution means relatively low toxicity.

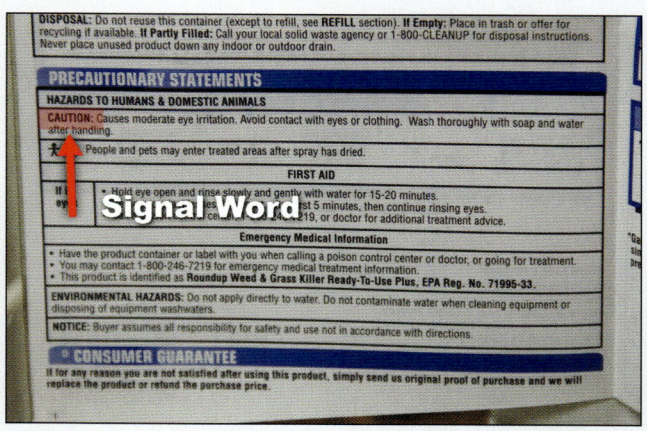

Step 3

- The label may also contain a *precautionary statement* such as keep from waterways or keep out of reach of children.

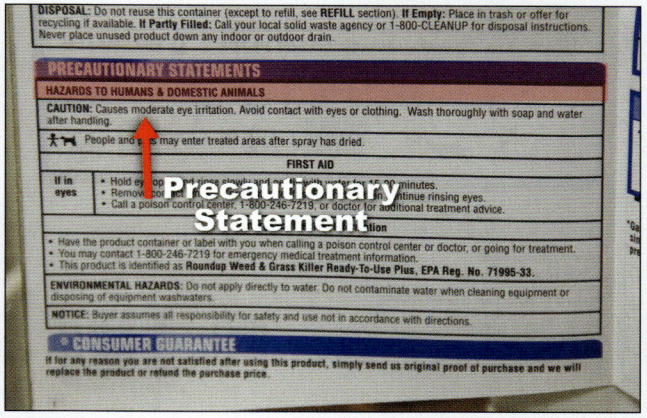

Step 4

- *Hazard statements* list special flammability, explosion or chemical hazards as well as if it is an environmental hazard. For example: Extremely flammable—flash point below 80°F.

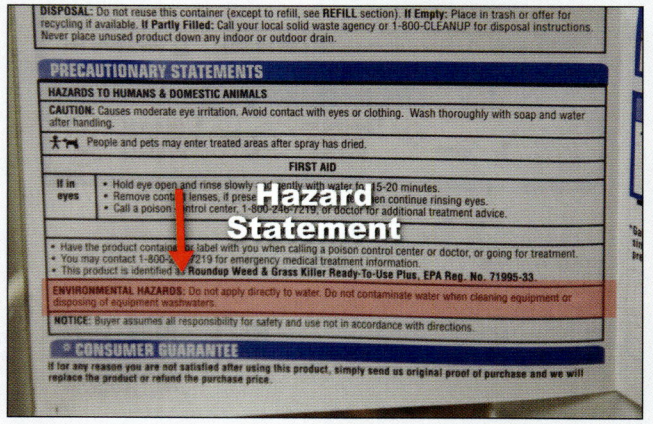

Step 5

- The *active ingredient* section lists the names and percentage of the active chemicals that make up the pesticide. Inert ingredients are listed as a percentage only.

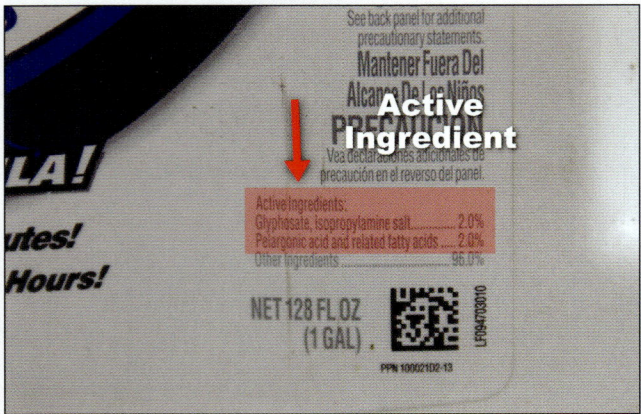

Step 6

- The *Environmental Protection Agency (EPA) registration number* can be found on the label. The EPA regulates the manufacturing and labeling of pesticides and requires a registration number on the label. Additional information can be obtained by contacting CHEMTREC and providing them with this number. The Canadian version of this number is the Pest Control Products (PCP) number.

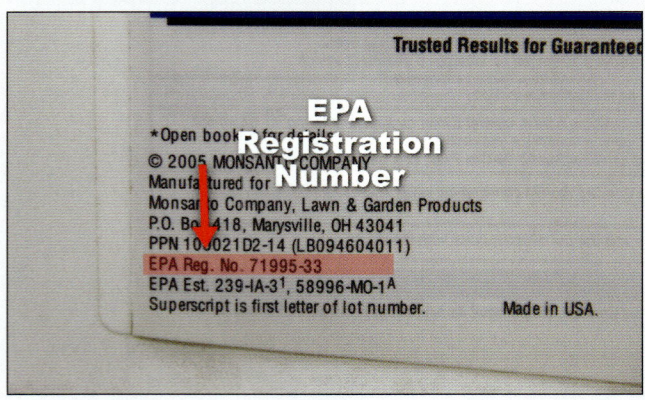

Step 7

- The label may also contain information on the mode of entry into the body, storage and disposal requirements, first aid information, and if known, an antidote in the event of poisoning.

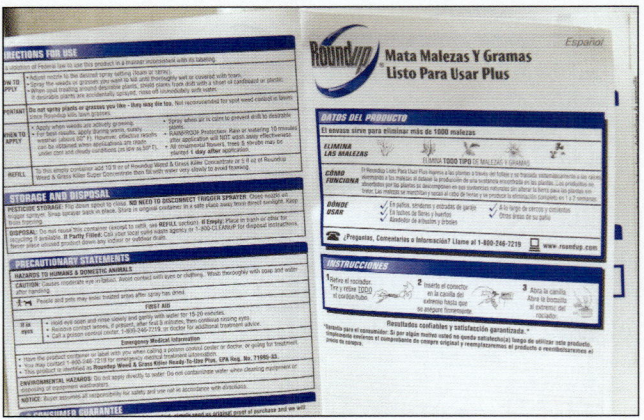

Using the Emergency Response Guide (ERG)

Every first responder needs to understand the Emergency Response Guide, or ERG. This booklet gives appropriate response guidelines to follow during initial operations.

Always approach a potential hazmat incident upwind.

Step 1

- First, from a distance, attempt to identify the material. This can be done by finding the four-digit ID number on a placard, an orange panel, shipping paper, or the package. Sometimes the name of the material is listed right on the package or shipping documents.

Step 2

- Next, identify which three-digit guide number corresponds to your specific substance. If you have the ID number, look up the three-digit code in the yellow-bordered pages of the ERG. If you know the name of the material, use the blue-bordered pages of the ERG.

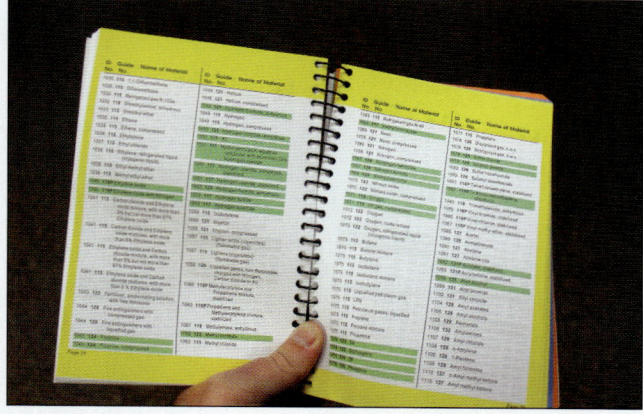

Step 3

- Substances that are highlighted in green are toxic inhalation hazard (TIX) substances. Use the green-bordered pages to identify the isolation and protective action distance that is required for that substance. Begin the isolation and protective action immediately.

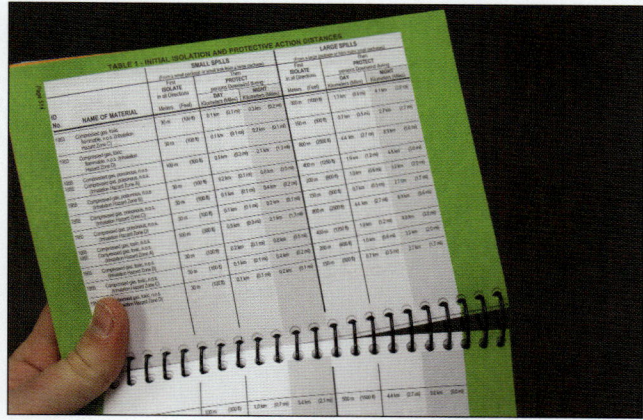

Step 4

- Now turn to the associated three-digit guide, which is located in the orange-bordered pages. Read this information very carefully.

Step 5

- If an associated placard is visible on the vessel, use the listed three-digit guide number on the Table of Placards and Initial Response Guide to Use On-Scene to proceed until more complete information is obtained.

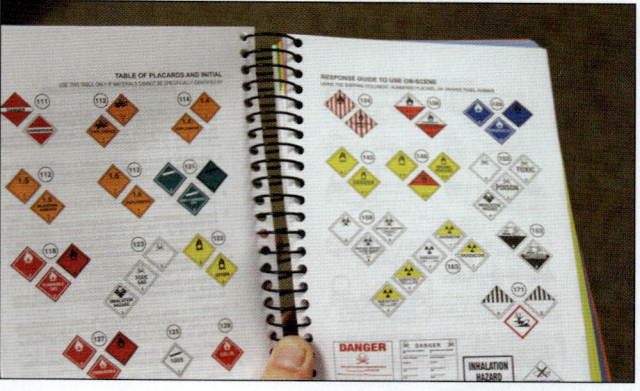

Step 6

- In the event that you are unable to find associated placards, the shape and style of the tank can give clues to its contents. This is a last resort. Turn to The Rail Car and Road Trailer Identification Chart to identify the rail car or trailer type. Use the listed three-digit guide number to proceed until more complete information is obtained.

FIREFIGHTER I

SKILL DRILLS

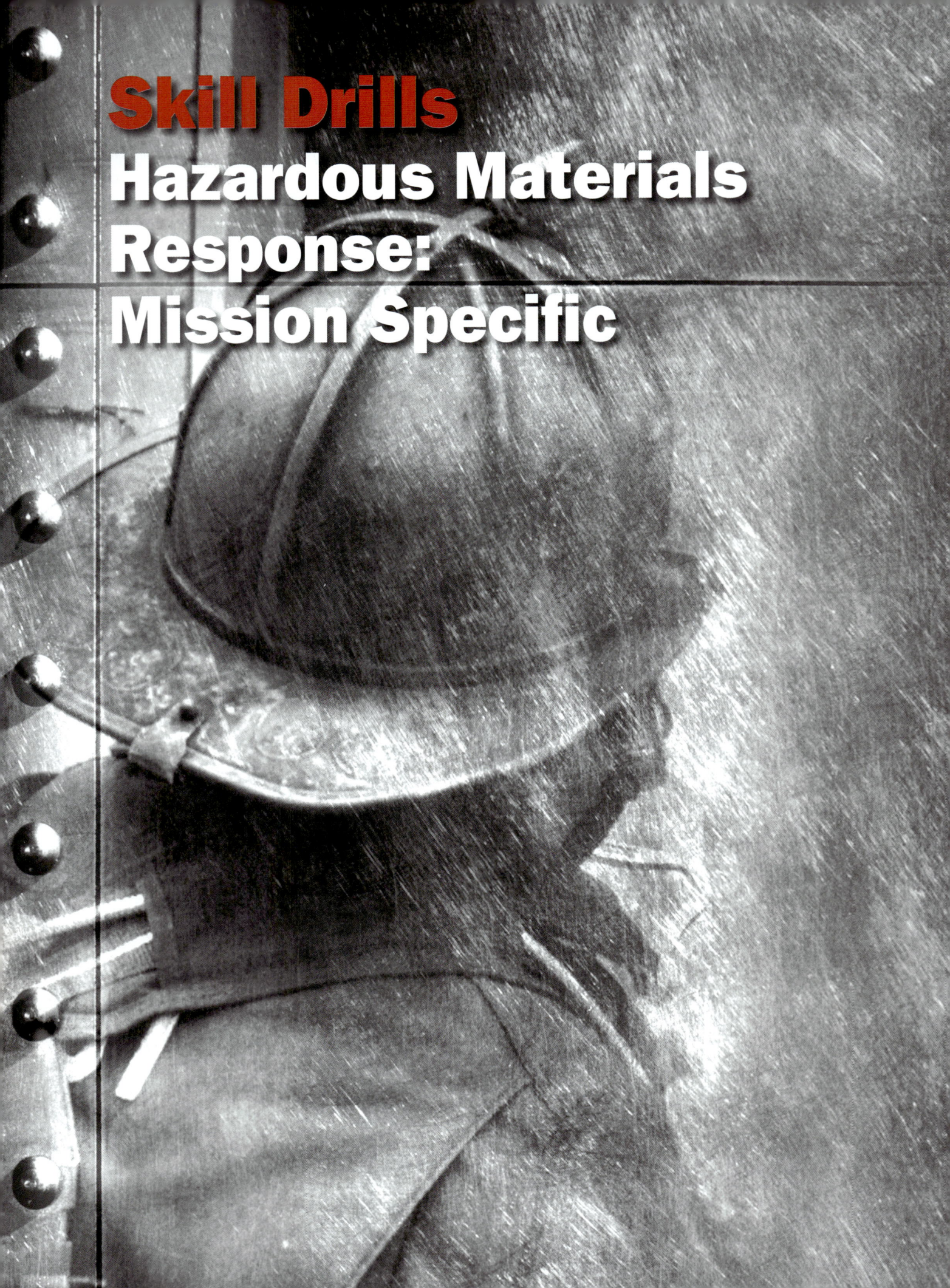

Skill Drills

Hazardous Materials Response: Mission Specific

These photos are for instruction only. Always follow your department's SOPs. When you perform these techniques at actual incidents, be sure to wear full protective gear and SCBA.

Care of Hazmat PPE

Step 1

- At regular intervals, hazmat gear must be inspected and tested to ensure it is in working order.

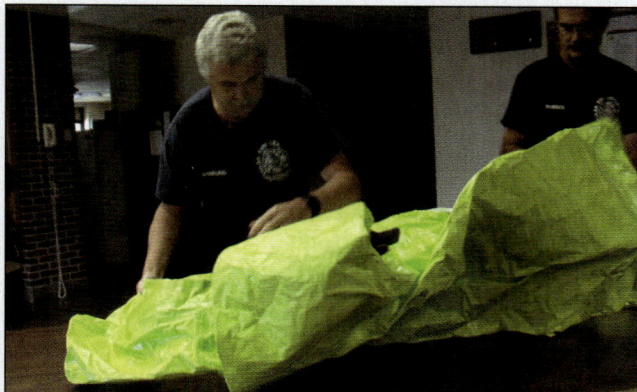

Step 2

- After removing the suit from the apparatus or storage area, lay it out on a clean surface. If it is a class A suit, lay it face down.

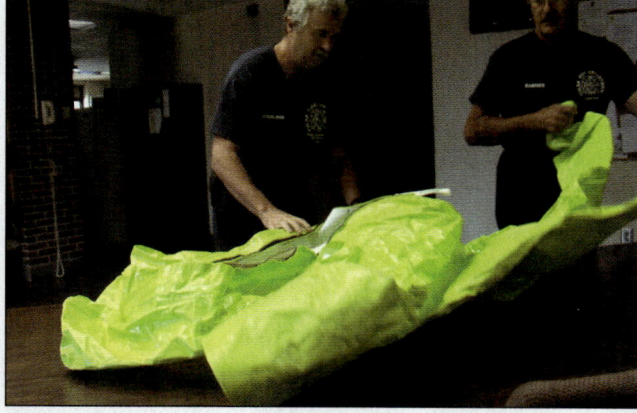

Step 3

- Carefully inspect the suit for damage or wear.

Step 4

- If it is a class A suit, it should be tested to ensure that it properly seals. Each manufacturer will have

its own testing equipment and instructions. Follow them carefully.

Step 5

- After attaching the testing equipment to the suit, pressurize the suit to the designated pressure.

Step 6

- Close the inlet supply to the suit and monitor the pressure gauge to ensure that the suit is holding its pressure.

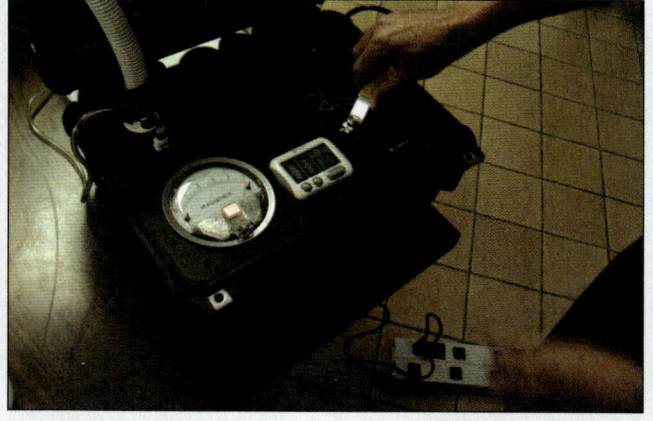

Step 7

- If the suit is losing pressure, use a spray bottle with a soapy water solution to spray on the suit and pinpoint the leak. Make sure to check gloves, zippers, foot areas, face masks, as well as the air inlet connections.

4 — Hazardous Materials Response: Mission Specific

FIREFIGHTER I

SKILL DRILLS

Step 8

- If the suit is leaking, it must be taken out of service and sent to an appropriate company to be fixed and recertified.

Step 9

- Once the suit is fully pressurized, start a stopwatch. The suit must hold a certain amount of pressure over a certain amount of time.

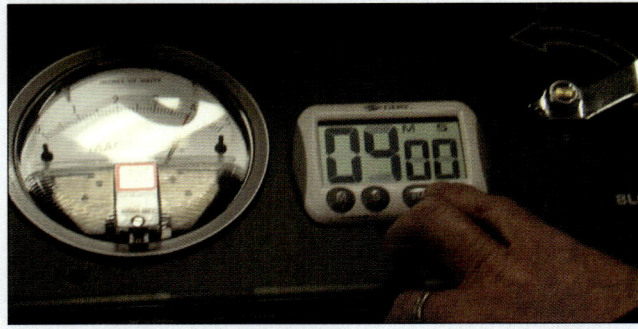

Step 10

- After passing the manufacturer's pressure test specifications, the suit should be refolded and placed in it's storage bag. The suit can now be placed back into service until its next inspection cycle.

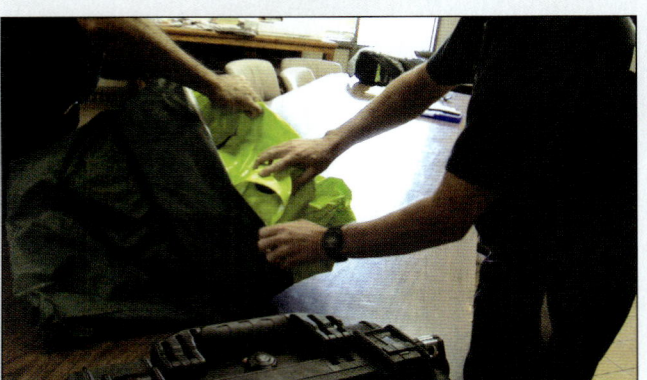

Step 11

- Ensure that the officer on duty properly documents the testing results.

Donning Hazmat PPE

When reporting to a hazmat incident, it is important to don the appropriate PPE. There are several types of suits available. Adjust this method to fit your response ensemble. Hazmat ensembles are often difficult to don and assistance is required.

Step 1

- After removing your boots, don the suit over your station uniform.

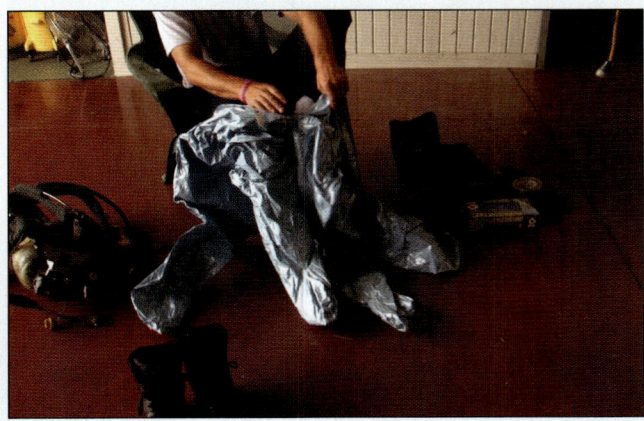

Step 2

- Zip up any enclosures.

Step 3

- Don the ensemble-approved boots over your suit. This will add additional protection to this area.

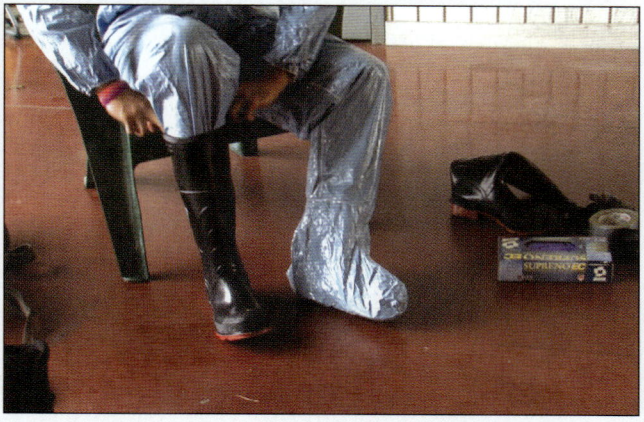

Step 4

- Don two sets of gloves for additional protection. The first set is a set of nitrile or latex gloves, depending on your department's preference. The outer gloves are ensemble-approved gloves.

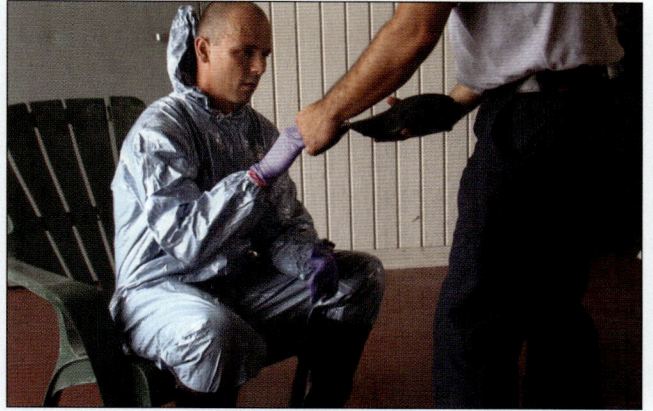

Step 5

- Using appropriate tape, cover the opening between the gloves and the ensemble. Also cover the opening between the top of the boots and the ensemble. When sealing up these sections, fold over the end of the tape. This will give you a place to tear away the tape during doffing of the suit.

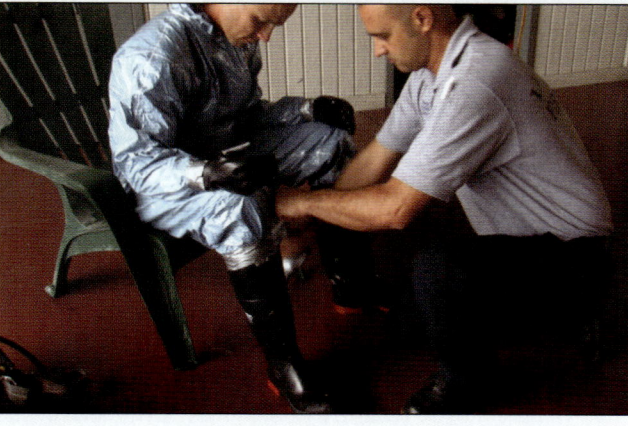

Step 6

- Now don the SCBA and mask.

Step 7

- Pull the ensemble hood over your head and overlap the edge of the SCBA mask.

Step 8

- Use appropriate tape to seal the gap between the SCBA mask and the suit.

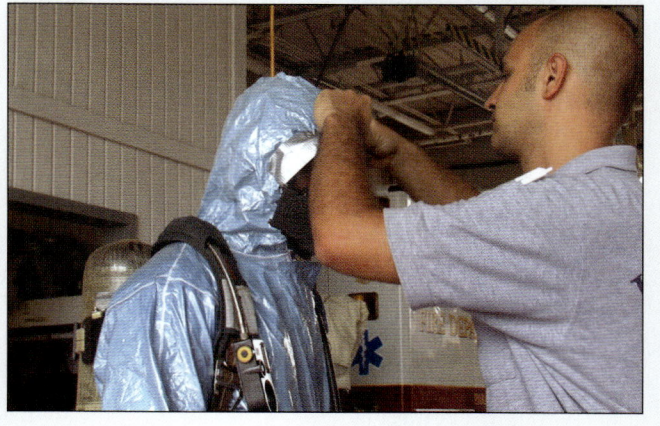

4 — Hazardous Materials Response: Mission Specific

FIREFIGHTER I

SKILL DRILLS

Step 9

- The firefighter can now insert a regulator and enter the hazmat zone.

Decontamination/Doffing of Hazmat PPE

- When exiting the hazmat zone, it is crucial to decontaminate the firefighter.

Step 1

- Have the firefighter stand in a basin to catch the water flow.

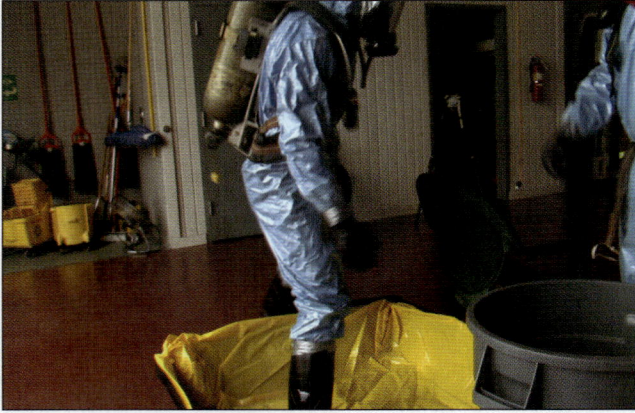

Step 2

- Rinse the firefighter with water.

Step 3

- Scrub the firefighter with a brush and an approved soap and do not touch the firefighter.

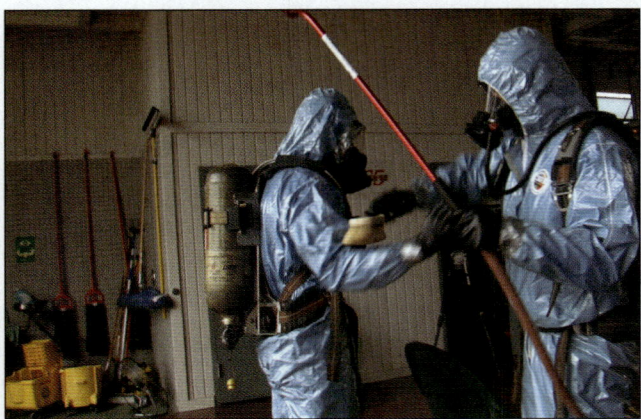

Step 4

- After thoroughly scrubbing and rinsing the firefighter, have him or her step out of the basin. Depending on the contaminants, some fire departments will use two rinse basins in sequence, scrubbing and rinsing the firefighter twice. They may possibly use a third basin to rinse the firefighter one final time.

Step 5

- The firefighter assisting with the decontamination now removes the sealant tape.

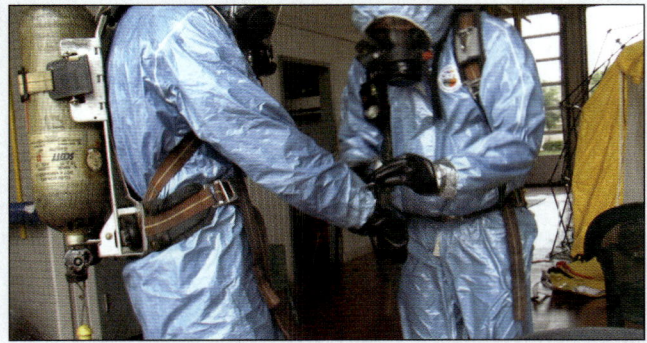

Step 6

- Pull back the hood of the hazmat suit.

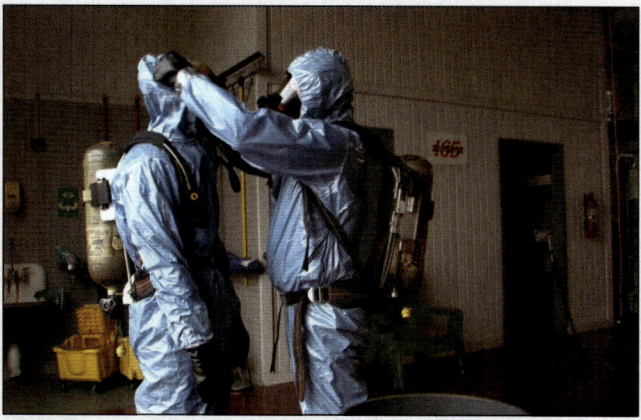

Step 7

- Remove the SCBA from the firefighter. Do not interrupt the airflow. Leave the facemask on the firefighter.

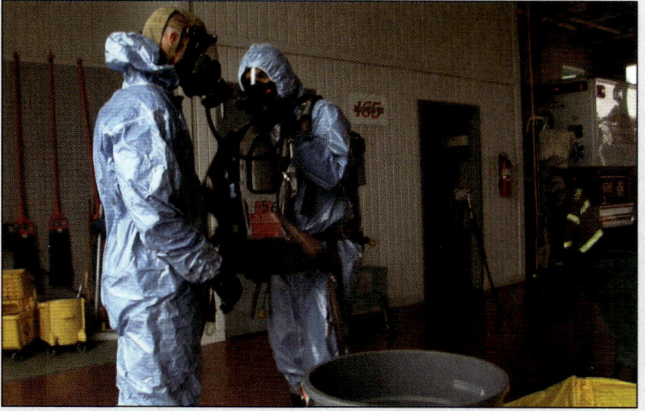

Step 8

- Now assist the firefighter in removing the outer gloves as well as the hazmat suit. As parts of the ensemble are removed, place them in the trash to be disposed of properly.

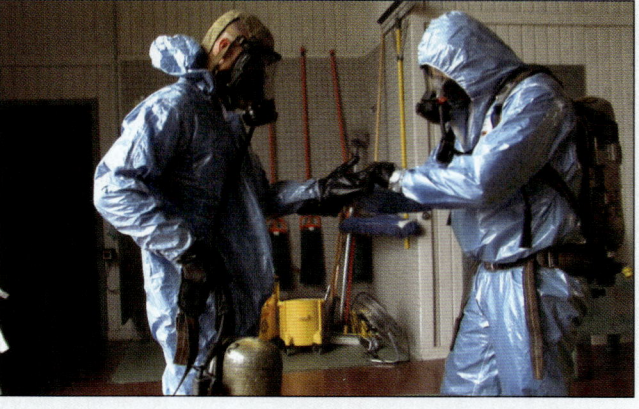

Step 9

- Assist the firefighter in removing the boots.

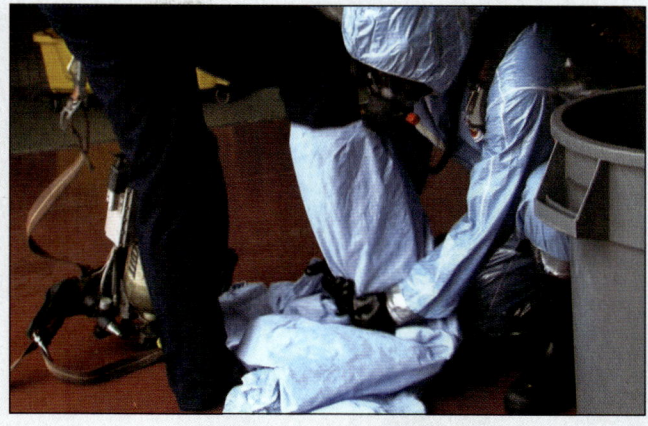

Step 10

- The firefighter can then remove interior gloves and facepiece.

Step 11

- Depending on the contamination, the firefighter may then need to be transported to a decontamination facility for fine decontamination.

4 — Hazardous Materials Response: Mission Specific

FIREFIGHTER I

SKILL DRILLS

Emergency Decontamination

Step 1

- Identify the victim requiring decontamination and move the victim to the decontamination area.

Step 2

- Assess the victim's air supply.

Step 3

- Release both shoulder straps and belt of the SCBA.

Step 4

- Remove the SCBA to the left shoulder side (the side the regulator hose is on to prevent wrapping the hose around the victim's throat). Maintain the victim's air supply. Do not remove the SCBA mask.

Step 5

- Place the SCBA on a chair and have the victim place one hand on the SCBA for stability.

Step 6

- Cut the hood from front to back over the head and down the center. You may need to remove the sealant tape in order to complete this step.

Step 7

- Cut the back of the suit from the previous cut. Cut down the center of the back and one leg to the top of the boot. Cut the top of the boot.

Step 8

- Cut down the other leg to the top of boot. Cut the top of the boot.

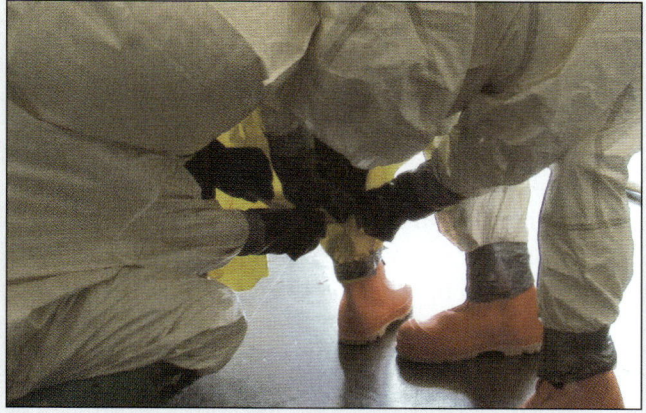

Step 9

- Loosen the outer gloves and pull over the victim's hands by grasping the individual fingertips of the outer gloves and pulling gently until the outer gloves slide free of hands. Arms and hands are still in the suit and the outer gloves are still taped to suit.

Step 10

- Grasp the front of the suit at the top and peel it down, removing the suit down to the boots (the victim will have to switch the hand stabilizing the SCBA during this step).

Step 11

- Step or hold the back of the boot and instruct the victim to step out of the boot onto the inside of the suit. Repeat with the other foot.

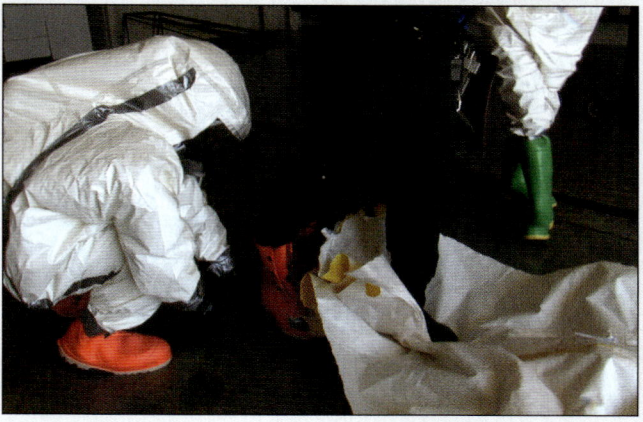

Step 12

- Flush the victim with copious amounts of water from top to bottom for a minimum of 30 seconds.

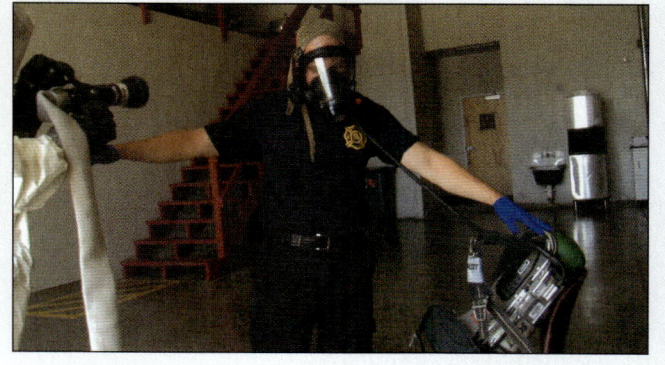

Step 13

- Remove all contaminated inner clothing without touching the victim with contaminated clothing. Leave the inner gloves on.

Step 14

- Flush the victim with copious amounts of water from top to bottom for a minimum of 30 seconds.

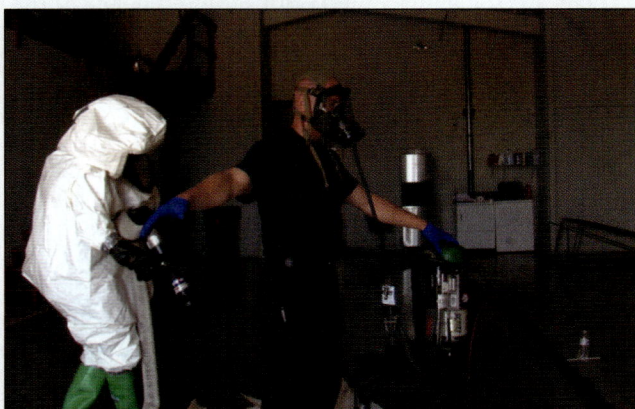

Step 15

- Remove the victim to a clean area. Carry the SCBA for the victim to a clean area.

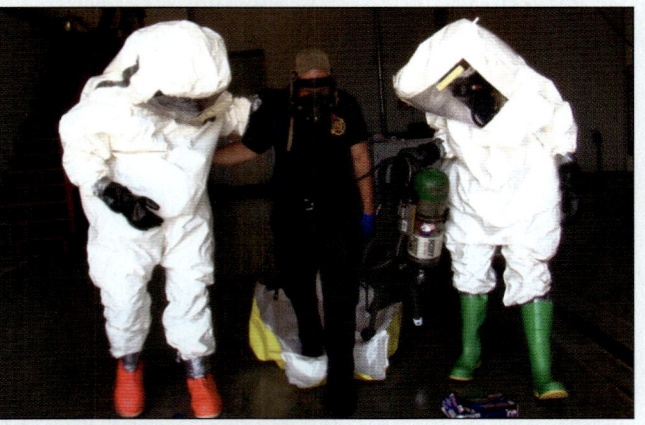

Step 16

- Have the victim grasp the mask under the chin and remove in one motion by pulling slightly forward and up and over the head. If possible, push the black button on the regulator to stop the flow of air from the SCBA.

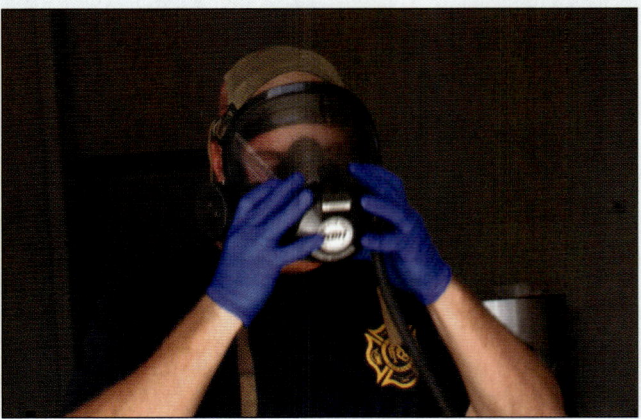

Step 17

- Remove the inner gloves by peeling them off while turning them inside out and folding them into one another to prevent touching the outside of the glove with a bare hand.

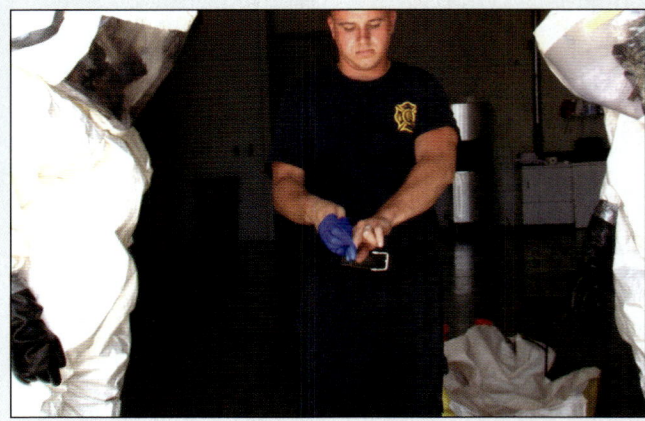

Step 18

- Transport the victim to the hospital.

Emergency Decontamination of a Civilian

Step 1

- Remove the victim from the contaminated area.

Step 2

- Wash the victim with flooding quantities of water.

Step 3

- Remove all contaminated clothing.

Step 4

- Continue to wash the victim.

Step 5

- Move the victim to an uncontaminated area.

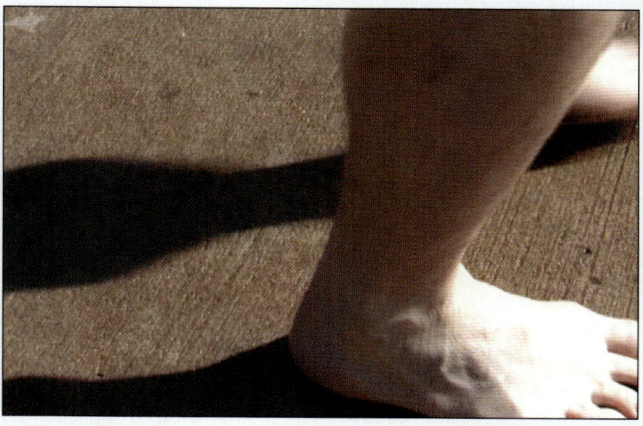

Step 6

- Begin first aid procedures and transport the victim to the hospital. Advise the ambulance and the hopital personnel of the contaminant involved.

Mass Decontamination

- Mass decontamination is used when a large number of people needs to be decontaminated to the best of our ability on a short notice. A mass decontamination setup can be as complex or simple as your equipment and time allows.

Step 1

- Take two apparatus and park them side-by-side, leaving a corridor in the middle.

Step 2

- Run a water supply to the apparatus. These operations take large amounts of water.

Step 3

- Place a fog nozzle on the output panel of each apparatus.

Step 4

- Set the nozzles on a wide fog pattern and supply them with a low flow. Open up the nozzles. You have now effectively created a decontamination corridor.

Step 5

- As the contaminated people approach, have them remove their clothing and walk down the corridor.

At the other end supply them with towels and robes if available.

Step 6

- When performing mass decontamination, consideration must be made for the water runoff. Often the sheer flow of water will dilute the substance to a safe level, but that is not always the case.

Step 7

- After mass decontamination, the people should be monitored and, depending on the contaminant, they may need to be transported to a facility for fine decontamination.

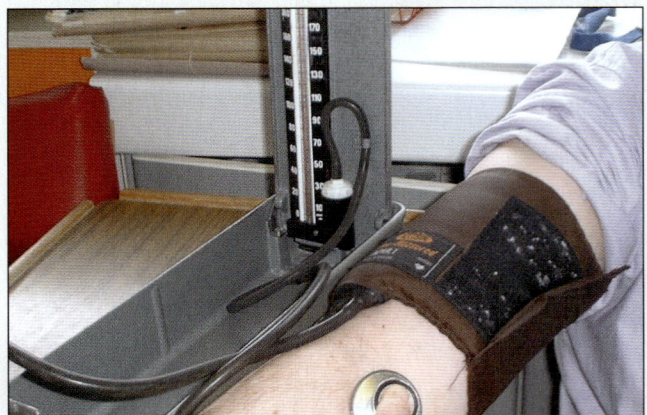

Absorption and Adsorption

Step 1

- Absorption is the physical process of "picking up" a liquid product in a spill. The liquid is trapped inside the absorption product.

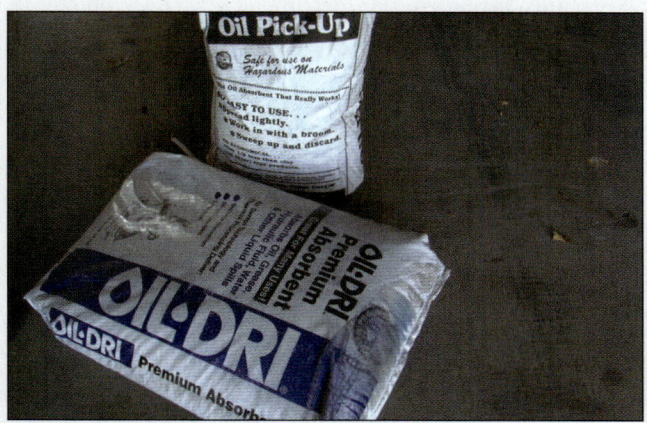

Step 2

- This can be a product as simple as kitty littler or speedy dry, used commonly at car wrecks to clean up oil spills.

Step 3

- Floating booms and pads are also examples of absorption devices.

Step 4

- Adsorption is the chemical process by which the spilled product adheres to the surface of the sorbent. There are quite a few examples of these as well.

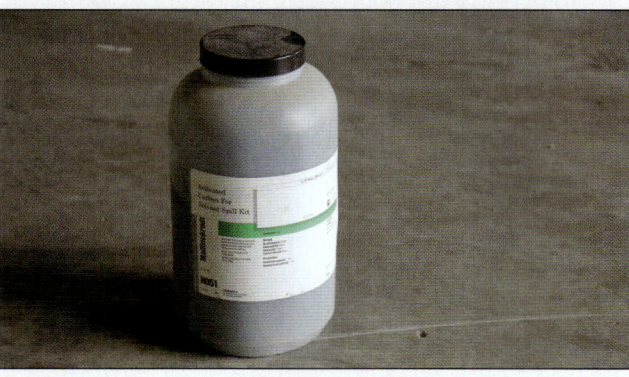

Step 5

- When using either and absorbent or adsorbent, the product must be disposed of in accordance with environmental regulations.

Damming

Step 1

- Damming is the process of constructing a barrier intended to slow or stop the flow of liquid into the environment.

Step 2

- It can be constructed of any material that could prevent a liquid from moving downhill. Dams can also be used as collection points.

Step 3

- Overflow and underflow dams are used when constructing a dam in moving water.

Diking and Diversion

Step 1

- To perform diking, use dirt or sand to construct a land-based barrier to control the movement of the liquid hazardous material.

Step 2

- Two common types are the V-dike, which collects the released material in the small end of the V, and the circle dike, which is typically a berm placed 360 degrees around a leaking container.

Step 4

- If the hazardous material has a specific gravity greater than 1, and is thus heavier than water, an overflow dam should be constructed. The hazardous material will drop to the bottom at the dam and the water, which is lighter than the hazardous material, will flow over the dam. Ensure that you use a pipe at least 4 inches in diameter.

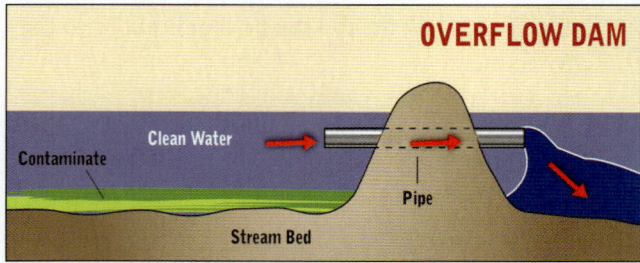

Step 5

- If the hazardous material has a specific gravity less than 1, and is thus lighter than water, an underflow dam should be constructed. The hazardous material will be trapped against the top of the underflow dam. The pipes at the bottom of the dam allow the water at the base of the dam to flow freely. Since most spilled hazardous materials float on water, this type of dam is most commonly used.

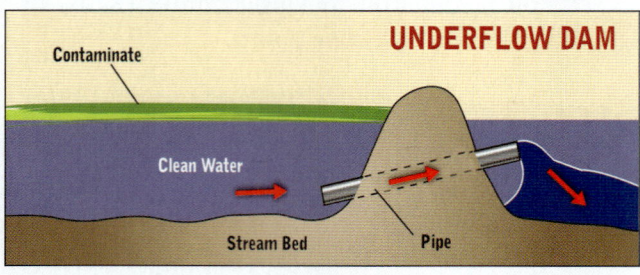

Step 3

- To perform diversion, place the dirt or sand in a position to channel the spilled materials into a containment area where it poses less harm.

Retention

Step 1

- Retention is the process of collecting the spill in a specific area.

Step 2

- When on water use an absorbent boom.

Step 3

- When dealing with a liquid on land, dams are used to create retention areas.

Step 4

- When dealing with a solid, use a tarp to cover the material.

Dilution

Step 1

- Dilution reduces the concentration of the material to a nonhazardous or less hazardous state.

Step 2

- This is typically done by diluting the substance with large quantities of water. This is not ideal when dealing with water reactive substances or when runoff is an issue.

Vapor Dispersion and Suppression

Step 1

- Vapor dispersion is moving gases or vapors to lower the concentration.

Step 2

- Vapor dispersion can be performed with a fog stream.

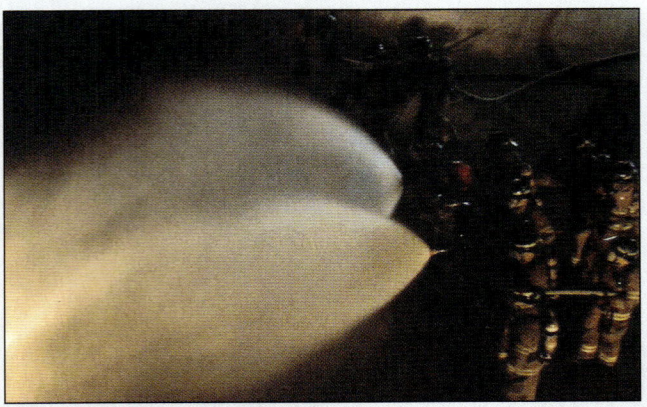

Step 3

- Vapor dispersion can also be performed with a fan.

Step 4

- Be sure to disperse the vapors into an area where they will not cause more harm.

Step 5

- Vapor suppression is the process of reducing or eliminating the vapors produced by a spill. Firefighting foams are the most common form of vapor suppression in use.

Step 6

- Choose the correct foam for your specific incident.

Step 7

- Apply the foam in an appropriate manner: rainfall technique, roll-on technique, or bank down technique.

Remote Shutoff

Step 1

- Emergency remote shutoff devices will vary from facility to facility. A remote shutoff valve may be manual, hydraulic, pneumatic, or mechanical.

Step 2

- It is important to first contact facility personnel to determine where the control for the valve may be and what the safest method is to accomplish the task. Do not open or close any valve without understanding its implications.

Step 3

- Contact the IC upon completing the task.

Highway Cargo Tank Remote Shutoff

Step 1

- Emergency shutoffs are usually well marked and located in easy-to-find areas. They are typically found behind the driver's side of the cab or near the control valves.

Step 2

- Here is the location of the emergency shutoff for an MC-306 trailer.

Step 3

- Here is the emergency shutoff for the MC-407.

Step 4

- Here is the emergency shutoffs for the MC-331 trailer.

Emergency Removal Using a Stretcher

Step 1

- In the event that a firefighter goes down during a hazmat situation, additional care should be given to avoid puncturing or breaching the hazmat suit.

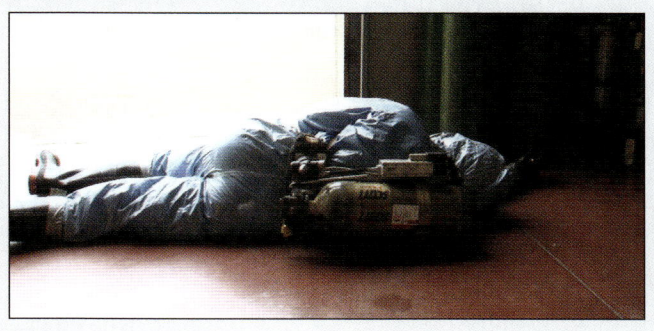

FIREFIGHTER I

SKILL DRILLS

Step 2

- Using a sked or other portable stretcher, roll or push the firefighter on top of the device.

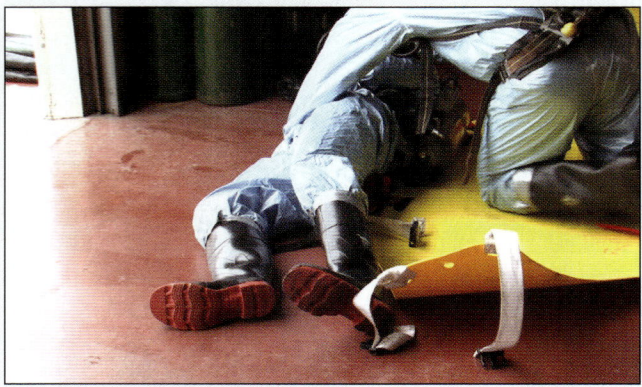

Step 3

- After securing the firefighter to the sked, simply drag it out of the hazard zone.

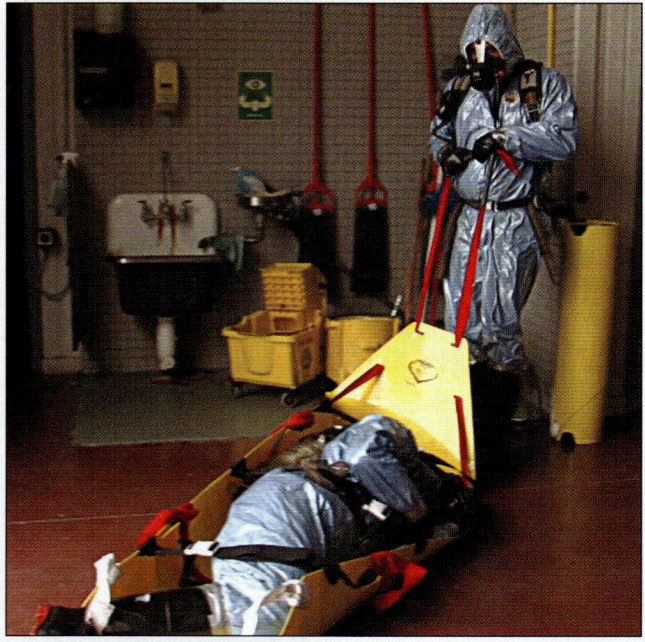

Pike Pole Removal

Step 1

- First locate the downed firefighter.

Step 2

- Loosen the firefighter's shoulder straps.

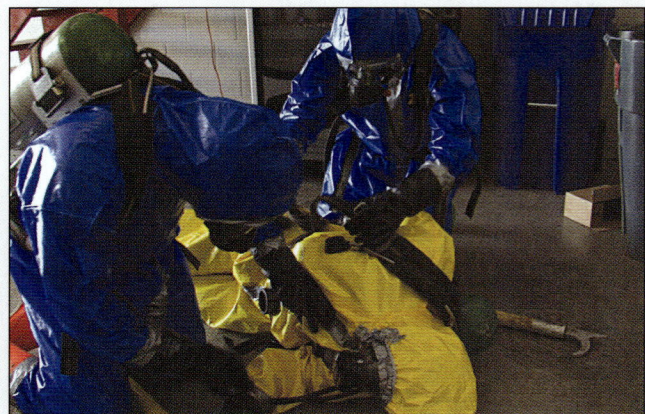

Step 3

- Place a pike pole through the shoulder strap. The end closest to the firefighter's feet should be on the inside of the shoulder strap.

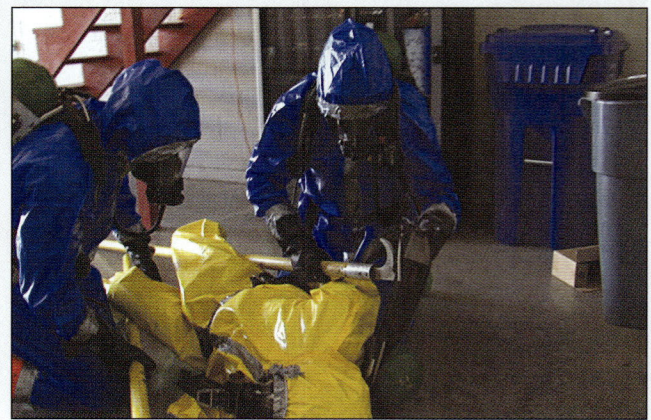

Step 4

- Tighten up the SCBA shoulder straps.

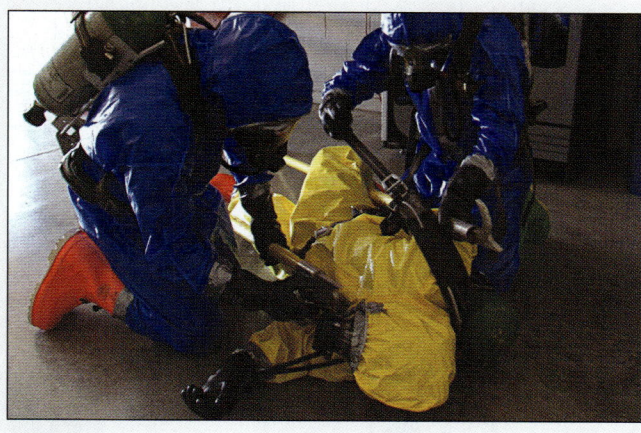

Step 5

- Drape the firefighter's leg over the pike pole, from inside to the outside.

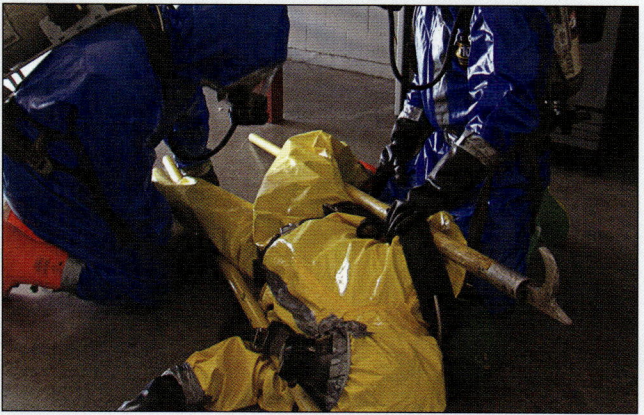

Step 6

- Do the same on the other side of the firefighter.

Step 7

- Positioning a firefighter at each end of the downed firefighter, lift up on the pike poles, lifting the firefighter up.

Step 8

- Remove the firefighter from the hazard zone.

Chair Removal

Step 1

- First locate the downed firefighter.

Step 2

- Place a chair next to the firefighter.

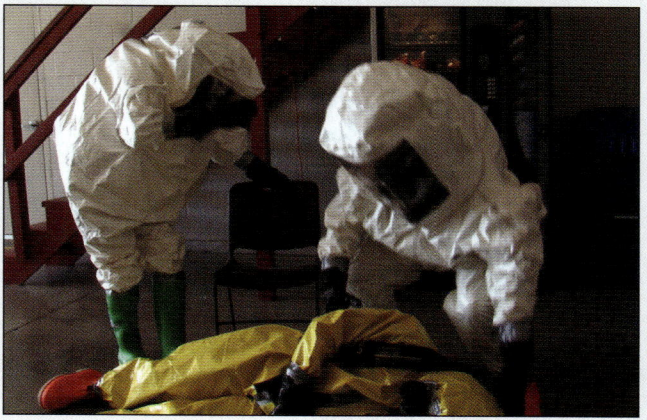

Step 3

- Loosen the SCBA shoulder straps.

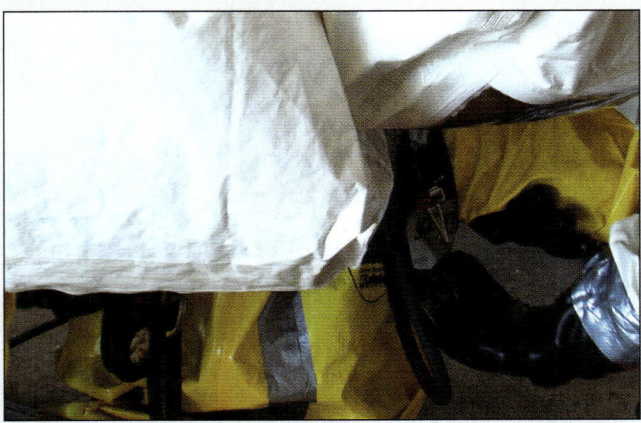

Step 4

- With one firefighter on each side, lift the firefighter up and into the chair.

Step 5

- Push the SCBA bottle to one side to allow the firefighter to sit securely in the chair.

Step 6

- With one firefighter on each side, lift the chair and remove the victim from the hazard zone.

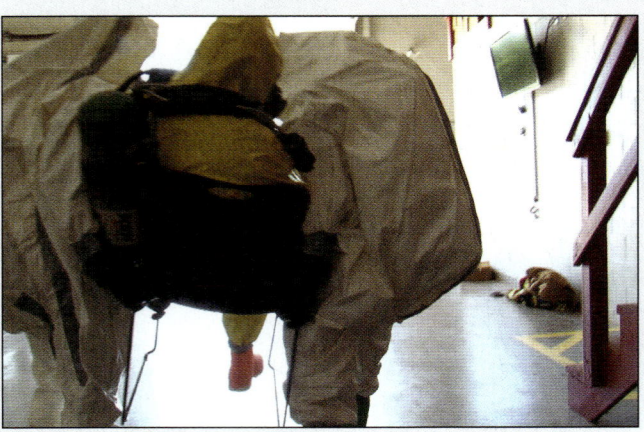

Incident Walkthrough

Step 1

- When responding to a potential hazmat incident, it is important to stay vigilant and follow all appropriate departmental SOPs.

Step 2

- When responding to a hazmat incident, always respond upwind, creating an appropriate control zone.

Step 3

- Staying an appropriate distance from the incident, analyze the scene and attempt to obtain any available information using binoculars to view the various markings on the container.

Step 4

- Using the ERG to identify the container, we can see that we have an intermodal tank. Now turn to the designated response guide, which in this case is 117.

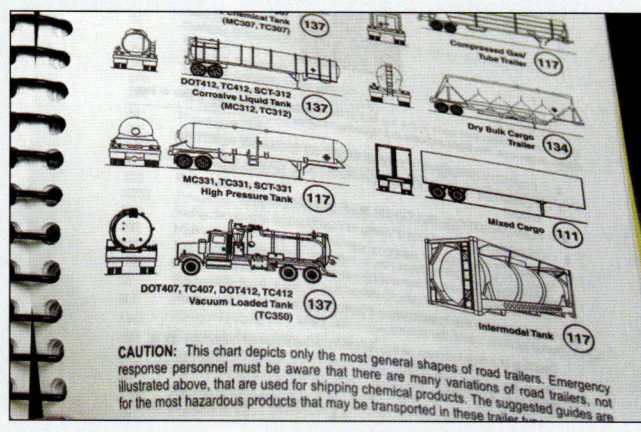

Step 5

- Guide 117 gives a blueprint for the response. During the incident, study this and use its guidelines as a minimum requirement. Guide 117 also designates an immediate precautionary isolation distance and provides a reference chart table in the green section of the book to look up specific distances once the chemical is identified.

Step 6

- During hazmat operations you must set up scene control zones and implement public protective actions as designated in the ERG. The police department is a great resource to assist in this task.

Step 7

- Prior to deploying personnel to the hot zone, it is important to plan out what the personnel are expected to do. In this case these personnel are being deployed to evaluate and mitigate the hazmat situation if possible.

Step 8

- As the evaluation continues, these firefighters are donning appropriate hazmat PPE and approaching the incident, preparing to implement the response consistent with the ERG.

Step 9

- Always work in teams of two. Two responders are fully suited up and preparing to enter the hazmat zone.

Step 10

- There are two firefighters on standby. These two firefighters are all suited up, except for donning their facemask and sealing up their suit. This team should not go on air until they are needed.

Step 11

- As the response team is directed by the officer to enter the hazard zone, they should carry appropriate monitoring equipment and check it regularly.

4 — Hazardous Materials Response: Mission Specific

FIREFIGHTER I

SKILL DRILLS

Step 12

- The response team maintains constant communication with the incident commander, communicating the status of the planned response at key points in the operation.

Step 15

- The markings on this vessel are inadequate, however the team has located the MSDS for this shipment inside the cab of the vehicle. This is one possible location for the MSDS.

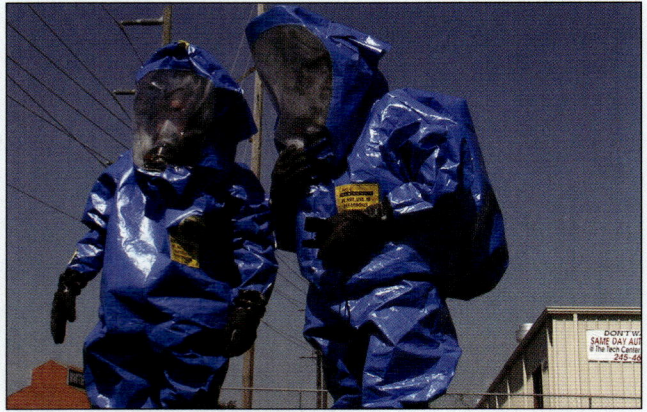

Step 13

- The team gathers information about the incident and relays it back to the incident commander. As more information is gathered, the incident commander can augment and adjust the response.

Step 14

- As the team approaches the incident, they should attempt to gather any additional information on scene, such as a bill of lading or any additional markings on the vessel.

Step 16

- This information should immediately be transferred to the IC, so the IC can adjust the response accordingly.

Step 17

- At this time, the team should attempt to close or contain the spill if possible. In this instance it was as simple as closing the valve.

Step 18

- Conduct any on-scene tests to verify the contents of the vessel and transfer that information to the IC.

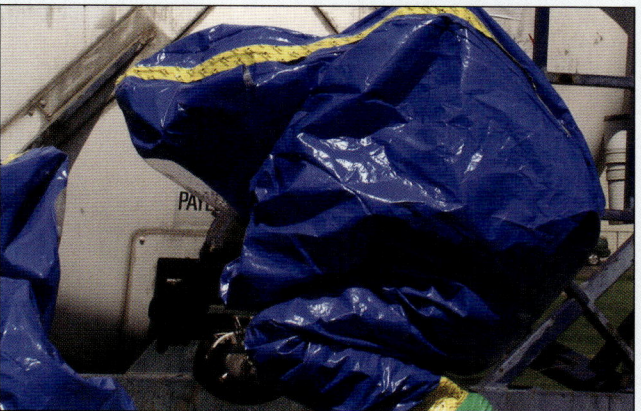

Step 19

- Make sure to monitor your air supply during the incident. Give yourself enough time to exit the hazard zone, go through the decontamination process, and remove the PPE prior to the low air alarm sounding.

Step 20

- The response team now goes through decontamination and reports to rehab to be checked by EMT personnel.

4 — Hazardous Materials Response: Mission Specific

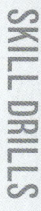

FIREFIGHTER I

SKILL DRILLS

Fire Engineering's Study Guide for Hazardous Materials Response

by Jason Emery

1. The minimum temperature at which a liquid gives off sufficient vapors to create an ignitable mixture with air near its surface is called the _____.

    a. boiling point

    b. ignition point

    c. flash point

    d. vapor point

2. Substances that have no definitive shape or volume are called _____.

    a. solid

    b. liquid

    c. gas

    d. vapor

3. Intermediate bulk containers (IBC) are commonly found in industry to transport product around an industrial site, they are usually square in shape and can hold up to _____ of liquid product or an equivalent amount of powders, slurries or other forms of product.

    a. 300 gal

    b. 330 gal

    c. 500 gal

    d. 555 gal

4. Most hydrocarbons have a specific gravity _____.

    a. less than 1

    b. greater than 1

    c. equal to 1

    d. Hydrocarbons are miscible.

5. _____ -level responders are trained to a greater level and operate using offensive tactics to control and stop the flow of the hazardous material at its source.

   a. Awareness

   b. Operations

   c. Technician

   d. Specialist

6. _____ materials ignite when they come in contact with each other.

   a. Pyrophoric

   b. Unstable

   c. Hypergolic

   d. Corrosive

7. A_____ has a pH value greater than 7.

   a. corrosive

   b. particle

   c. acid

   d. base

8. A_____ has a pH value less than 7.

   a. corrosive

   b. particle

   c. acid

   d. base

9. _____ is a term that refers to the size of a solid and usual measured in microns.

   a. Particle size

   b. Physical size

   c. Specific size

   d. Typical size

10. _____ is defined by the US Department of Transportation as any substance or material capable of posing an unreasonable risk to health, safety, and property when transported in commerce and which has been so designated.

   a. Hazardous substance

   b. Hazardous chemical

   c. Hazardous waste

   d. Hazardous material

11. The weight of a gas compared to the weight of air is known as _____.

   a. vapor pressure

   b. vapor density

   c. vapor gravity

   d. specific gravity

12. Also known as general service cars, these tank cars have a typical capacity of 4,00 to 45,000 gal and operating pressures between 30 to 100 psi.

   a. nonpressure tank car

   b. pressure tank car

   c. cryogenic tank car

   d. dry hopper car

13. _____ reactions occur with most chemical reactions and cause a release of heat.

   a. Endothermic

   b. Exothermic

   c. Hypergolic

   d. Pyrophoric

14. The DOT requires that hazmat loads of _____ or more be marked with the appropriate placard indicating the materials primary hazard.

   a. 1,000 pounds

   b. 1,001 pounds

   c. 1,010 pounds

   d. 1,100 pounds

15. UN 1993 is commonly displayed on many tanker trucks. This number is used when transporting.

   a. diesel

   b. anhydrous ammonia

   c. propane

   d. gasoline

16. The _____ of a material can also determine the most likely method of exposure.

   a. specific gravity

   b. particle size

   c. persistence

   d. physical state

17. Which NFPA Standard requires operations-level training be part of the Firefighter I Curriculum?

   a. NFPA 472

   b. NFPA 1001

   c. NFPA 1072

   d. NFPA 1991

18. These tanks contain gases or liquids under pressure starting at 100 psi. They have capacities ranging from 2,500 to 11,500 gal and have a bolted manway at the front or rear.

   a. DOT412/MC312

   b. DOT406/MC306

   c. DOT407/MC307

   d. MC331

19. These types of tank trucks are used commonly to transport gasoline, fuel oil, kerosene and other flammable and combustible liquids.

   a. DOT412/MC312

   b. DOT406/MC306

   c. DOT407/MC307

   d. MC331

20. A large network of pipelines runs throughout the United States. Like other transport methods these pipelines fall under the jurisdiction of the _____.

   a. OSHA

   b. NFPA

   c. DOT

   d. CFR

21. _____ is the ability of a material to undergo a chemical reaction with another substance.

   a. Chemical reactivity

   b. Specific reactivity

   c. Physical reactivity

   d. Catastrophic reactivity

22. The container _____ can provide valuable insight into the type of material contained inside.

   a. color

   b. size

   c. weight

   d. shape

23. The NFPA 704 marking system consist of color coded diamonds with a number system to indicate the level of danger for specific hazards. Which of the following statements is correct?

    a. The numbering system ranges from 1 to 4.

    b. The left diamond section (red) denotes the flammability of the material.

    c. The lower the number, the greater the hazard.

    d. This system does not identify the products.

24. The _____ was enacted to control airborne emissions to protect the environment.

    a. Clean Air Act

    b. Oil Pollution Act

    c. Resource Conservation and Recovery Act

    d. Environment Amendments and Amendments Act

25. Carboys, Dewar flasks, drums, compressed gas cylinders are types of _____ packaging.

    a. bulk

    b. non-bulk

    c. special

    d. excepted

26. The weight of a substance compared to the weight of equal volume of water is called _____.

    a. specific density

    b. specific gravity

    c. vapor density

    d. vapor pressure

27. A(n) _____ reaction is a chemical reaction that is accompanied by the absorption of heat.

    a. endothermic

    b. exothermic

    c. hypergolic

    d. pyrophoric

28. Substances that destroy or burn living tissues and have destructive effects on other type of materials are known as _____.

    a. corrosive

    b. persistent

    c. unstable

    d. erosive

29. The ability of a liquid to mix with water is called _____.

    a. solubility

    b. adsorption

    c. insolubility

    d. dilution

30. Caution, Warning, and Danger are examples of _____.

    a. package hazard words

    b. signal words

    c. guide words

    d. label marking words

31. Substances that have no definite shape but do have a specific volume are called _____.

    a. solid

    b. liquid

    c. gas

    d. vapor

32. Hazardous materials are divided into _____ hazard classes

    a. 7

    b. 8

    c. 9

    d. 10

33. _____ materials are those that react and ignite upon coming in contact with air.

    a. Exothermic

    b. Unstable

    c. Hypergolic

    d. Pyrophoric

34. NFPA standard _____ is a marking system used for fixed bulk storage tanks and vessels in fixed facilities

    a. 472

    b. 704

    c. 1001

    d. 1500

35. Good _____ skills play a key role on the initial response and size-up.

    a. management

    b. tactical

    c. observation

    d. firefighting

36. _____ is the temperature at which a liquid changes to a gas at normal atmospheric pressure.

    a. Flash point

    b. Vapor point

    c. Ignition point

    d. Boiling point

37. The pressure at which a liquid and its vapor are in equilibrium at a given temperature is known as _____.

    a. critical pressure

    b. boiling pressure

    c. vapor pressure

    d. water pressure

38. The minimum temperature that a fuel in air must be heated to in order to ignite without an independent ignition source being present is called the _____.

    a. flash point

    b. ignition temperature

    c. vapor point

    d. boiling point

39. A substance that has both a specific shape and volume is called a _____.

    a. solid

    b. liquid

    c. gas

    d. vapor

40. The percentage of the gas or vapor concentration in air that will burn if an ignition source is introduced is known as the _____.

    a. flash range

    b. flammable range

    c. lower explosive limit

    d. upper explosive limit

41. Concentrations between the upper and lower explosive limits will burn rapidly if ignited, and any concentrations above _____ of the LEL must be considered to have potential for ignition.

     a. 5%

     b. 10%

     c. 25%

     d. 50%

42. Energy released in the form of particles or electromagnetic waves is called _____.

     a. explosive energy

     b. radiation

     c. thermal energy

     d. nuclear energy

43. There are _____ types of radiation.

     a. 5

     b. 4

     c. 3

     d. 2

44. _____ radiation has enough energy to cause a physical change to atoms by triggering them to lose electrons and become ions.

     a. Acute

     b. Chronic

     c. Ionizing

     d. Nonionizing

45. _____ are relatively heavy and slow moving and can be stopped by a piece of paper or skin.

     a. Alpha particles

     b. Beta particles

     c. Gamma rays

     d. Neutrons

46. _____ are fast and can move at speeds of 9/10 the speed of light.

     a. Alpha particles

     b. Beta particles

     c. Gamma rays

     d. Neutrons

47. Gamma rays are _____ rays that move at the speed of light and can kill living cells.

    a. light

    b. proton

    c. electromagnetic

    d. nonionizing

48. _____ consists of penetrating particles and is much harder to shield against, and primarily results from the fissioning or splitting of certain atoms within a nuclear reactor.

    a. Alpha radiation

    b. Beta radiation

    c. Gamma radiation

    d. Neutron radiation

49. _____ radiation has enough energy to cause molecules to move or vibrate, but not enough to change them chemically.

    a. Alpha

    b. Beta

    c. Nonionizing

    d. Proton

50. _____ substances can be transmitted between a person or animal to another through some manner of contact.

    a. Contagious

    b. Infectious

    c. Thermal

    d. Biometric

51. _____ will present their symptoms immediately.

    a. Acute exposures

    b. Chronic exposures

    c. Acute effects

    d. Chronic effects

52. A large dose in a short period of time is known as _____.

    a. acute exposure

    b. chronic exposure

    c. acute effect

    d. chronic effect

53. _____ can often have both short-term and long-term effects, depending on the substance involved.

   a. Acute effects

   b. Acute exposures

   c. Chronic exposures

   d. Chronic effects

54. Heavy perspiration, skin moist to the touch, physical weakness, and muscle cramps are signs and symptoms of _____.

   a. heat cramps

   b. heat exhaustion

   c. heatstroke

   d. dehydration

55. An exposure to a microorganism that could result in a disease is known as a _____.

   a. biological exposure

   b. etiological exposure

   c. mutagen exposure

   d. teratogen exposure

56. Inert gases that displace the oxygen required for breathing are called _____.

   a. convulsants

   b. simple asphyxiants

   c. chemical asphyxiants

   d. carcinogens

57. Substances that prevent the body from using oxygen are called _____.

   a. convulsants

   b. simple asphyxiants

   c. chemical asphyxiants

   d. carcinogens

58. _____ are materials that are known to cause or are suspected of causing cancer in people.

   a. Asphyxiants

   b. Convulsants

   c. Toxins

   d. Carcinogens

59. Chemicals that have destructive effects on materials and burn or destroy living tissue are called _____.

    a. carcinogens

    b. convulsants

    c. toxins

    d. corrosives

60. There are signs and symptoms of a chemical exposure that by their name affect a specific organ rather than the entire body. _____ affect the kidneys.

    a. Neurotoxins

    b. Hepatoxins

    c. Nephrotoxins

    d. Etiotoxins

61. What is the most common route of exposure?

    a. ingestion

    b. inhalation

    c. injection

    d. absorption

62. The maximum exposure time, limited to 15 minutes no more than four times a day without suffering harmful effects is called _____.

    a. immediately dangerous to life and health (IDLH)

    b. threshold limit value/ceiling (TLV-C)

    c. threshold limit value/time weighted average (TLV/TWA)

    d. threshold limit value/short term limit value (TLV/STEL)

63. The most common and widely used resource by emergency responders at hazardous materials/WMD incidents is _____.

    a. the state and federal emergency hotline

    b. the Emergency Response Guidebook

    c. industry representatives

    d. local hazmat teams

64. The Emergency Response Guidebook is updated and released every _____ years.

    a. 2

    b. 3

    c. 4

    d. 5

65. The _____ colored pages of the ERG list the UN/NA ID numbers in numerical order.

    a. white

    b. orange

    c. blue

    d. yellow

66. For which mode of transportation would you find shipping papers titled "Dangerous Cargo Manifest"?

    a. highway

    b. rail

    c. water

    d. air

67. The injury or damage resulting from exposure to the hazardous materials release and subsequent exposure is called _____.

    a. contact

    b. harm

    c. injury

    d. damage

68. The area that is adjacent to and downwind of the spill and in imminent danger of becoming contaminated within 30 minutes of the materials release is the _____.

    a. initial isolation zone

    b. protective action distance

    c. potential harm area

    d. immediate endangered area

69. The ability of an atmospheric monitor to provide accurate readings must be checked on a regularly scheduled basis. This check is referred to as _____.

    a. bump testing

    b. relative response testing

    c. calibration

    d. response time adjustment

70. Responders using atmospheric monitors must take into consideration the physical state and chemical properties of a product. Because of a product's _____, a responder holding an atmospheric monitor at waist level may not obtain accurate readings.

    a. vapor density

    b. specific gravity

    c. vapor pressure

    d. specific density

71. Everyone not directly involved in emergency response operations must be kept away from the area. This _____ task is done to establish control over the area of operation

    a. evacuation

    b. protective

    c. priority

    d. isolation

72. _____ should be conducted when the incident involves uncontrollable leaks, explosive materials, or unknown gas leaks from large-capacity containers.

    a. Isolation

    b. Protective action

    c. Evacuation

    d. Establishing control zones

73. During a hazardous materials incident, the assistant safety officer-hazardous materials reports directly to the _____.

    a. incident commander

    b. hazardous materials group supervisor

    c. hazardous materials branch director

    d. incident safety officer

74. The first responder trained to the operations level should be familiar with available resources such as hazmat response teams, mutual aid fire, police and EMS agencies etc. A list of these available resources should be established, maintained and included with the resource listing in the _____.

    a. LERP

    b. LEPC

    c. command post

    d. MSDS binder

75. When assessing the risk involved with making a rescue, the responder should consider the general properties for each of the hazard classes. At a minimum, before performing rescue operations be sure to determine the _____.

    a. shape of the container

    b. density of the material

    c. specific hazard of the material

    d. specific gravity of the material

76. Confinement is a(n) _____ action in which the flow of the hazardous material is controlled.

    a. offensive

    b. defensive

    c. protective

    d. technical

77. Containment is a(n) _____ action that prevents additional releases from a container.

    a. offensive

    b. defensive

    c. protective

    d. technical

78. The _____ is the most commonalty used breathing apparatus in the fire service.

    a. open-circuit SCBA

    b. closed-circuit SBCA

    c. supplied air respirator

    d. air purifying respirator

79. The difference between a level B suit and a level C suit is _____.

    a. hand protection

    b. splash protection

    c. respiratory protection

    d. communications

80. The ability of a garment to resist damage from direct contact with a chemical is _____.

    a. chemical resistance

    b. permeation

    c. penetration

    d. degradation

81. The process by which a hazardous chemical moves through a given material at the molecular level is _____.

    a. chemical resistance

    b. permeation

    c. penetration

    d. degradation

82. The flow of a hazardous chemical through openings in the material such as zippers, seams, pinholes or other imperfections is _____.

    a. chemical resistance

    b. permeation

    c. penetration

    d. degradation

83. The physical destruction or breakdown of the material after being exposed to a chemical is _____.

    a. chemical resistance

    b. permeation

    c. penetration

    d. degradation

84. _____ incidents are characterized by a rapid onset of medical symptoms which may take minutes or hours.

    a. Chemical

    b. Biological

    c. Radiological

    d. Etiological

85. In _____ incidents the onset of symptoms may take days to weeks.

    a. chemical

    b. biological

    c. radiological

    d. etiological

86. _____ agents are most likely to be used in a terrorist incident because they are the easiest for a terrorist to produce.

    a. Chemical

    b. Biological

    c. Radiological

    d. Etiological

87. _____ is a widely available toxin that is easily produced and is ten thousand times more toxic than sarin.

    a. Anthrax

    b. Mycotoxins

    c. Tularemia

    d. Ricin

88.   Radioactive agents fall under hazard class _____.

    a. 5

    b. 6

    c. 7

    d. 8

89.   The Federal Bureau of Investigation Bomb Data Center says that incendiary devices are used in _____ of all bombing incidents in the United States.

    a. 5% to 10%

    b. 10% to 15%

    c. 15% to 20%

    d. 20% to 25%

90.   Tabun, Sarin, Soman, and V-agent are all examples of _____ agents.

    a. blister

    b. nerve

    c. blood

    d. choking

91.   Mustard, lewisite, distilled mustard, and nitrogen mustard are examples of _____ agents.

    a. blister

    b. nerve

    c. blood

    d. choking

92.   The initial _____ should evaluate what has occurred thus far as well as asses current conditions and what is likely to occur.

    a. incident action plan

    b. safety briefing

    c. safety plan

    d. evaluation plan

93.   OSHA _____ requires states that employees who are expected to participate in emergency responses should be given training in hazardous materials.

    a. 1910.120

    b. 1910.134

    c. 1910.472

    d. 1910.1200

94. The physical process of picking up a liquid product spill is called _____.

    a. adsorption

    b. absorption

    c. dispersion

    d. dilution

95. _____ reduces the concentration of the material to a non hazardous or less hazardous state.

    a. Adsorption

    b. Absorption

    c. Dispersion

    d. Dilution

96. The use of water spray to direct hazardous vapors away from a certain area is called _____.

    a. adsorption

    b. absorption

    c. dispersion

    d. dilution

97. _____ is the transfer of a hazardous material out of the hot zone in quantities greater than those deemed acceptable.

    a. Contamination

    b. Dilution

    c. Suppression

    d. Dispersion

98. _____ is the reduction or the elimination of vapors produced by spilled hazardous materials.

    a. Dispersion

    b. Suppression

    c. Dilution

    d. Decontamination

99. There are _____ types of emergency shut off valves.

    a. 1

    b. 2

    c. 3

    d. 4

100. _____ is the process that occurs when people, equipment, or the environment come in contact with a hazardous material.

   a. Contamination

   b. Exposure

   c. Contagious

   d. Infectious

CHAPTERS 1–6 ANSWER KEY

Question #	Answer	Page #	Question #	Answer	Page #
1	c	55	33	d	23
2	c	54	34	b	27
3	c	67	35	c	46
4	a	54	36	d	55
5	c	80	37	c	55
6	c	53	38	b	55
7	d	54	39	a	54
8	c	54	40	b	56
9	a	54	41	b	56
10	d	7	42	b	56
11	b	55	43	d	56
12	a	65	44	c	56
13	b	714	45	a	56
14	b	17	46	b	56
15	a	21	47	c	56
16	d	47	48	d	56
17	b	5	49	c	56
18	d	64	50	a	57
19	b	63	51	c	57
20	c	28	52	a	57
21	a	53	53	b	57
22	d	63	54	a	57
23	d	15	55	b	58–59
24	a	6	56	b	60
25	b	68–69	57	c	60
26	b	54	58	d	60
27	a	53	59	d	60
28	a	54	60	c	61–62
29	a	55	61	b	62
30	b	26	62	d	62
31	b	54	63	b	31
32	c	22	64	c	31

CHAPTERS 1–6 ANSWER KEY *Continued*

Question #	Answer	Page #	Question #	Answer	Page #
65	d	32	97	a	108
66	c	41	98	b	123
67	b	52	99	c	127
68	b	39	100	b	109
69	c	76			
70	a	77			
71	d	42			
72	c	82			
73	d	89			
74	a	42			
75	c	81			
76	b	81–82			
77	a	77			
78	a	98			
79	c	106			
80	a	105			
81	b	105–106			
82	c	106			
83	d	106			
84	a	151			
85	b	151–152			
86	b	153			
87	d	153			
88	c	153			
89	d	154			
90	b	154			
91	a	155			
92	b	107			
93	a	1			
94	b	62			
95	d	110			
96	c	109			

INDEX